Directory of Short-Run Book Printers

Third Edition

An international directory of book printers capable of producing from 10 to 10,000 copies of a book, booklet, catalog, magazine, journal, manual, directory, or other bound publication

Compiled and written by
John Kremer

AD-LIB PUBLICATIONS

51 N. Fifth Street
P.O. Box 1102
Fairfield, IA 52556-1102

SAN 265-170X   ISBN Prefix 0-912411
Telephone (515) 472-6617

Directory of Short-Run Book Printers, Third Edition

compiled and written by John Kremer

Published by: Ad-Lib Publications
               51 N. Fifth Street
               P. O. Box 1102
               Fairfield, IA 52556-1102

               (515) 472-6617

Third Edition

Printed and bound by Walsworth Publishing
                     Marceline, Missouri

Cover photo provided by Publishers Press
                        Salt Lake City, Utah

Library of Congress Cataloging in Publication Data

Kremer, John, 1949-
    Directory of short-run book printers.

    Bibliography: p. 232
    Includes index.
    1. Printers--United States--Directories.
2. Book industries and trade--United States--
Directories
Z475.K73 1985   686.2'025'73   84-12313
ISBN 0-912411-06-6 (pbk.)

Table of Contents

Other books published by Ad-Lib Publications:

FormAides for Direct Response Marketing, Rev. Ed.
  by Ad-Lib Consultants
ISBN 0-912411-02-3 / LC 83-21497
80 pages (8 1/2 x 11) perfectbound / $9.95

"If you have any questions about the nitty-gritty of
mail order, you can find the answer here. ... The
material is short, concise, information-packed, and
easy to find. An excellent resource."
                                    -- The Professional Quilter

FormAides for Successful Book Publishing
  by Ad-Lib Consultants
ISBN 0-012411-00-7 / LC 83-060510
52 pages (8 1/2 x 11) saddlestitched / $5.95

"It taught me instantly how to save dozens of hours in
the routine office work of a small publisher."
 -- George Moberg, author/publisher, Writing in Groups

The Independent Publisher's Bookshelf, Third Edition
  by John Kremer
ISBN 0-912411-07-4
32 pages (8 1/2 x 11) saddlestitched / $3.00

"As a soon-to-be publisher, I marvel at the wealth of
information. As a writer, I enjoyed the tight,
sprightly delivered prose."
                                    -- Dan Rapoport, independent publisher

Coming in the spring of 1986, two new books:

Book Marketing Made Easier
  by John Kremer
ISBN 0-912411-11-2
96 pages (8 1/2 x 11) perfectbound / $10.00

101 Ways to Market Your Books -- and Make Money
  (For Publishers and Authors)
  by John Kremer
ISBN 0-912411-08-2 hardcover / $19.95
ISBN 0-912411-09-0 softcover / $14.95
approximately 200 pages

How to Use This Book

This _Directory_ is designed to save you time and money in finding the best printer for your next book, catalog, journal, magazine, manual, annual report, directory, or other bound publication. It could save you hundreds, even thousands of dollars in printing and binding costs alone. Plus, of course, it can save you time and hassles.

As one user of the Second Edition wrote: "Without your _Directory_ I ended up paying $12,500 for the first printing of one book -- and I got plenty of production hassles. With the _Directory_, I paid $4,500 for the identical job -- and not one problem."

How can you get these savings? You need to do two things: (1) contact the appropriate printers, and (2) query them in a professional manner. This _Directory_ will help you with both steps.

First, its listings of almost 400 book printers will help you to select the appropriate one for your job. Read these listings carefully. Check to see which printers can offer the services you require. Can they print the quantity you want? Do they offer the type of binding you need? What about their other services, their credit terms, their reputation for quality and dependability?

Then, once you have found five or ten book printers who you think could do the job you require, query them. Always query at least five printers. Don't get hooked on using one printer all the time. As your needs change, you will find that you may also have to change printers to find one who can handle the different quantities, sizes, or bindings that you require ... with the quality you want ... at a price you can afford ... and with great service and dependability.

To query the printers, use your letterhead stationery. Type in the specifications for your book, make copies, and send a copy to each printer.

For more about querying printers, read the section on How to Request a Printing Quotation.

OTHER FEATURES OF THIS NEW THIRD EDITION

This Third Edition of the Directory of Short-Run Book Printers has been expanded to list over 200 more book printers than the Second Edition, including 11 Canadian and 30 Asian and European printers, for a total of almost 400 printers.

Many of the listings provide detailed information on the capabilities and services of each printer:

* the minimum, maximum, and optimum quantities they can handle;
* their standard book sizes;
* what bindings they can offer in-house;
* what other bound publications they can handle (magazines, journals, galley copies, catalogs, annual reports, juvenile picture books, cookbooks, school yearbooks, and general commercial printing);
* what other services they offer (typesetting, teletypesetting, design, editing, 4-color printing, warehousing);
* the payment terms they offer small publishers;
* and many other details and comments.

We've also added the results of a user survey we conducted in the spring of 1985. This survey rates over 50 printers on their speed, price, dependability, service, quality, and overall capability. The survey results for the individual printers are included as part of their listings.

Another new section (which users of the Second Edition insisted we add) is the listing of printers classified by areas of specialization. This edition includes classified lists for 16 different categories:

* ultra-short run
* short run
* self-publishing
* typesetting
* 4-color
* hardcovers in house
* annual reports
* catalogs
* cookbooks
* demand printing
* galley copies
* journals
* magazines
* mass-market paperbacks
* picture books (juvenile)
* yearbooks

Plus a state by state listing of all the printers included in this edition of the Directory so you can easily locate those book printers nearest you.

And for those of you who love exotic climes, we've listed 11 Canadian book printers and 30 overseas printers in separate sections. The section on overseas printers also includes tips on how to deal most effectively with such printers.

But that's not all. We've also included two of our most popular reports, "17 Points to Consider When Selecting a Book Printer" and "20 Ways to Save Money on the Printing of Your Books" (now expanded to 38 ways!). We have been selling these reports for $2.00 each. Now they are free as part of this Third Edition because we feel these points are essential to getting the most out of this Directory.

We've also written another new report, "17 Ways to Save by Printing Short Runs," and included it in this edition as well.

And, finally, we've once again included the most revealing section of our last edition -- the price comparison charts that let you compare actual price quotes we've received from printers for the production of some of our books. These charts will open your eyes.

For this current edition, we've added three new charts sent to us by other publishers. Here's an example of what they reveal: a difference of over $12,000 between the highest and lowest bids for 5000 copies of a 160-page, 8 1/2 x 11 photo book printed on glossy paper. That's $2.40 per book in savings. And you say it doesn't pay to query?

Besides this regular book edition of the Directory, we now offer a Deluxe Mail-Merge Edition on floppy disc for your convenience in querying printers. This Deluxe Edition includes an already formatted RFQ form, all the lists in the classified section (again, with names and addresses already formatted for mail merging), and label formats for one, two, or three across label printing.

This Deluxe Edition sells for $30.00 ($25.00 if you have already bought this book edition). It is available for a large number of computers. See the section, Mail Merge Deluxe Edition, for more information on ordering this special edition.

Ad-Lib Publications
P. O. Box 1102
Fairfield, IA 52556-1102

SAN 265-170X

(515) 472-6617

From: John Kremer

REQUEST FOR QUOTATION   Please quote by June 30, 1985

Quote your best price and delivery in producing the following:

Specifications

Title: Directory of Short-Run Book Printers, 3rd Edition

Total pages: 240

Trim size: 5 1/2 x 8 1/2 inches

Text paper: 55 lb. book paper

Text ink: black

Cover stock: 10 pt C1S / varnished

Cover ink: 4-color sides 1 and 4 / separations provided

Binding: perfectbound, with three color inserts

Copy: camera-ready line copy
               some already screened copy to be provided

Packing: in tightly sealed cartons

QUOTE

Quantity: 5000 - $_____

Shipment will be _____ working days from receipt of copy.

Terms: Net 30 (credit references available upon request).

Remarks:

Printer _____   _____
Contact _____   Signature
Address _____
         _____   Phone: _____

Thank you. We look forward to doing business with you.

HOW TO REQUEST A PRINTING QUOTATION

To obtain a reliable quote for the printing of your book, you should supply all the information the printer will need to make a valid estimate. To ensure that you have included all the necessary information -- and to ensure that the same basic data is used by all printers you query -- use a Request for Quotation (RFQ) form on your letterhead. See page 8 for a copy of the RFQ we sent out when requesting quotations for Second Edition of this <u>Directory</u>.

You do not need to copy the form we used; you may adapt it to your own requirements. Regardless of what kind of format you decide to use, you should provide the printer with the following basic information (they need all of the following information in order to make an accurate estimate):

1) the **title** of the book --
   If you do not have a title yet, give some reference title or number.

2) the **number of copies** you want printed --
   You may list several options if you are not sure
   how many copies you want printed, but don't ask
   for more than two (just as a common courtesy).

3) the **number of pages** in the book --
   Include all pages: title page, copyright page,
   index, blank pages, and so on. Note that most
   printers can give you a better price if the
   number of pages in your book is a multiple of 32
   pages (96, 128, 160, 192, 224, 256, ...).

4) the **trim size** of the book --
   Standard trim sizes (such 5 1/2 x 8 1/2, 6 x 9,
   or 8 1/2 x 11) will enable you to get a better
   price from many printers <u>and</u> will fit the standard shelving units of libraries and booksellers,
   but don't rule out an odd size if it is appropriate for the contents of your book.

5) **text paper** --
   Most books are printed on 50 lb. or 60 lb. white
   offset or book paper, but if you are publishing a
   children's or photo book, you will probably want
   to use a different paper stock. If you want the
   book to last a hundred years or more, be sure to
   specify acid-free paper (which is, however, more
   expensive).

6) **text ink** --
   If you do not specify a color of ink, it will be black. If you want a four-color book or another accent color, be sure to specify it.

7) **binding** --
   Do you want your book to be a perfectbound softcover (like this book), or a saddlestitched book (stapled like a magazine), or a smyth-sewn casebound (hardcover), or comb or spiral bound (like many cookbooks)?

8) **cover stock** --
   Most trade paperbacks use a 10 pt. C1S (coated one side) cover, though other cover stocks are available. I recommend you ask for a varnish or, better yet, a film lamination coating for the cover (to better protect its surface during shipping and handling; also, it looks better).

9) **cover ink** --
   Will the cover be printed with one, two, or more colors? Any dropouts or screens? Be sure to also specify whether you want the cover printed sides 1 and 4 only (the outside) or all sides (including the inside).

10) **copy** --
    Will you provide camera-ready copy, or will you require typesetting and pasteup services (if you require typesetting, you need to provide them with a close estimate of the number of words and pages in the book, type size and font, etc.)?

    Will the copy have any photos? Bleeds? Extensive solid areas? If the book is to be printed in four colors, will you be providing the color separations or transparencies?

11) **packing** --
    Do you want your books shrinkwrapped singly or in convenient multiples (5 to 10 books) to protect them during shipment? Do you have any other special packing or shipping requirements?

An RFQ form lays out all this necessary information in a clear, understandable format so that any printer should know exactly what you want. If you do not know how to specify all the details regarding paper stocks, bindings, and so on, then either (1) read some books on graphics arts and printing, like <u>Publish Yourself without Killing Yourself</u> by L. A. Tattan or <u>The</u>

Graphic Designer's Handbook by Alastair Campbell (see the Bibliography in the back of this book), or (2) find a local graphic designer who can help you, or (3) ask questions of the printers you are considering using.

Several printers issue publications to help you understand all the terminology and will send these to you free if you write on your letterhead requesting a copy. Here are several of the best:

D. Armstrong's Books, Budgets, & Authors

Braun-Brumfield's Book Manufacturing Glossary, Book Paper Samples, and Type Sample Book

Delta Lithograph's Planning Guide

Dinner & Klein's Catalogs, Brochures, and Flyers

R. R. Donnelley's Guide to Book Planning

Friesen Printers' Book Publishers' Guide

McNaughton & Gunn's Book Manufacturing Intro Kit

Thomson-Shore's superb Printer's Ink newsletter

MORE NOTES ON RFQ'S

On your RFQ form, besides providing the specifications for your book, you should also require a few other details from the printer. For instance, always get their credit terms, normal delivery times, and the approximate delivery charges to your firm (if any).

Normal terms of credit are net 30 with approved credit. In other words, if your credit is good, you will be expected to pay for the books 30 days after the books are shipped. If your credit is shaky, or your business is new, then most printers will require at least 1/3 down, another 1/3 with returned proofs, and the balance on delivery.

Delivery times can vary from as short as 10 working days (2 weeks) to as long as 8 to 12 weeks for camera-ready copy. If your book requires typesetting or case binding, the time can vary from as short as 4 weeks to as long as 20 or more weeks. If you need fast turnaround, let them know that when you query them.

When choosing a printer, be sure to consider more than just price. How about their turnaround time, their quality, their service, their terms? Read the next section, "17 Points to Consider When Selecting a Book Printer," to make sure you've considered all the facts before committing yourself to a printer.

A Few Final Points on Querying Printers

** Always ask the printer to quote by a specific date (allow 2 - 3 weeks minimum). Printers say they can produce a quote in 1 - 7 days and they should be able to, but our experience is that most printers take much longer. By setting a cutoff date, you ensure that all serious bids will be sent to you in time for you to make your decision.

** When you've narrowed your choice to two or three printers, ask them to send you samples of their work. Be sure to inspect these samples carefully. Also ask them to give you the names and phone numbers of some of their recent customers. Call these customers and ask them for feedback regarding the printer's quality and service. Do this before you send your money or camera-ready copy to a printer. You do not want to spend thousands of dollars for books you would not be proud to sell.

** When you are ready to send in your manuscript or camera-ready copy, phone the printer beforehand to confirm prices and the printing schedule. Prices may change due to changes in paper prices, and delivery schedules vary with the printer's workload. Be sure to confirm both. When you do send in your copy, enclose a written letter of confirmation (or contract) regarding the agreed-upon price and delivery schedule. This will save you from any doubt if some question should come up later regarding prices, services, or delivery.

** Always keep a copy of any manuscripts or camera-ready copy you send them. The postal service and printers have both been known to lose or spoil even the best prepared and packaged copy.

Note: Be sure to let the printers know you read about them in this Directory. It will show them that you are serious about getting good quality work for a reasonable price.

17 POINTS TO CONSIDER WHEN SELECTING A BOOK PRINTER

1. **Experience** -- How long have they been printing books? How long have they been in business? What kind of books are they accustomed to producing?

2. **Reputation** -- Do they have a reputation for doing quality work and delivering on schedule? Do you know anyone who has used them before? Can they provide references?

3. **Quality of Work** -- Do they produce good-looking, well-made books? Do their books hold together? Is the printing clear -- neither overinked nor under-inked? Have you seen samples of their work? Be sure to get a sample of their work before you commit to them -- preferably a sample similar in size and binding to the book you want produced.

4. **Price** -- How do their prices compare to other book printers? If their prices are higher than other printers, what added benefits do they offer? Faster delivery? Better quality? Greater reliability? More services?

5. **Service** -- Are they willing to work with you to produce the best book for the price you are able to pay? Do they answer your phone calls, letters, and other queries quickly and courteously? Do they put forth that extra bit of effort that makes working with them a pleasure?

6. **Dependability** -- Do they deliver books, catalogs, or other publications when promised? Do they live up to their promises -- both verbal and written?

7. **Timing** -- Do they offer faster delivery than other printers? Most book printers offer a 4 to 6 week lead time for books printed from camera-ready copy (2 to 3 weeks for catalogs and some other bound publications). Can this printer offer a faster delivery time and yet produce a quality book?

8. **Terms** -- What kind of terms can they offer you? Do they require a large downpayment? Do they offer any discounts for prepayment, or quick payment?

9. **Other Services** -- Can they provide warehousing and fulfillment services for you? Can they do typesetting and pasteup in house? Do they have teletypesetting capabilities?

10. **Quantities** -- What quantities are they capable of producing? Will they be able to follow up an initial short-run with a much larger second run, or will you have to go to another printer to do a higher quantity when (and if) your book begins to sell?

11. **Capabilities** -- What kind of books are they accustomed to doing? What sizes and types of bindings? Is your book a special size or binding? Will it economically fit their presses or binding capabilities.

12. **Specialization** -- Whenever possible, use a printer who specializes in the type of book you want produced. Find out which sizes, quantities, bindings, etcetera, they are accustomed to doing. Can they do odd sizes? And, if so, are they set up for easy handling of such odd sizes or bindings?

13. **Choices** -- Does the printer offer a choice of paper and cover stocks? Do they keep them in stock, or will they have to special order them? Special orders will usually require more time.

14. **Location** -- Where are they located? Location may be important for several reasons: 1) You may be able to get speedier delivery -- both in getting copy to them and in getting books from them; 2) It will be easier to make changes in the galleys if necessary; and 3) If you live nearby, you may have a chance to visit their plant and see more samples of their work and meet the people you will be working with.

15. **Equipment** -- Is the printer's equipment up to date? Can it handle the type of book you want produced? Is the equipment messy and dirty, or well maintained? (If they do not care how their plant or equipment looks, do you think they will care what your book looks like?)

16. **Packing and Shipping** -- Will the printer pack the books properly so that they are not damaged in shipment? Can they provide shrinkwrapping of books? Can they ship to more than one destination? Can they ship by both UPS and truck? Do they have the capability to handle your list maintenance and fulfillment (for journals and other periodicals)?

17. **Fitness** -- Be sure to select a printer who is right for the job you want done. Don't go to a quick printer for a casebound book, or to a web printer for 100 copies of a short brochure.

17 WAYS TO SAVE BY PRINTING SHORT RUNS

1) **Keep initial costs down** -- Obviously your total costs will be lower if you print 2000 copies of a book rather than 10,000. This lower initial cost can be important if you are working with limited funds.

2) **Print only what you need** -- Why print more copies than you can actually sell or more than you need to distribute? For example, if you have a membership of 800, why print more than 800 copies of your directory?

3) **Cut inventory and storage costs** -- Less inventory means less storage space is needed; hence, less rent. Plus, in states with inventory taxes, less inventory means less taxes paid.

4) **Get reprints to market quicker** -- Quicker response to market conditions will usually result in more sales. And most printers can produce a shorter run faster than they can a longer run.

5) **Test market books before moving to full production** -- Testing can save you from making too heavy a commitment to a title that proves, in actual market tests, not to be as big a seller as you had projected.

6) **Sell books with a limited market** -- Short runs allow you to add titles that will sell to specialized markets with a limited but affluent or information-hungry audience.

7) **Reduce the number of remainders** -- Any books sold as remainders lose you money (even when you sell at your production costs) because overhead and other hidden costs are rarely covered in a remainder sale.

8) **Free up money for promotion** -- Shorter runs allow more money to be freed up for promotion. And promotion (advertising, publicity, interviews, etc.) is what sells books.

9) **Update more frequently** -- You can provide up-to-the-minute information by updating your directories, manuals, or catalogs with every new print run.

10) **Risk less on experimental or uncertain titles** -- Shorter runs keep costs down so you don't tie up your working capital on uncertain titles.

11) **Keep your money costs down** -- A minimum inventory of books means that less of your capital is at risk which, in turn, means that your money costs (borrowing and opportunity costs) are less.

12) **Maintain a backlist without undue inventory** -- You can reprint backlist titles in short runs to take advantage of continuing sales to old and new customers. A larger backlist will often result in higher volume direct sales to bookstores and libraries.

13) **Market limited editions of worthwhile books** -- Shorter runs, with their attendant lower intial costs, allow you to publish books you would normally not be able to afford to publish.

14) **Prepare prepublication review copies** -- Send review copies (reader's proofs or galley copies) to important review media and opinion makers far in advance of publication. Also send copies to experts and professionals whose testimonial you would like to obtain for the book.

15) **Avoid obsolescence** -- Why get stuck with too large a supply of books or catalogs which are out of date?

16) **Cut down on mistakes** -- Short runs allow you to correct mistakes in prior editions. You can make your books or catalogs better and better with each reprinting.

17) **Eliminate the need to forecast long-term sales** -- Forecasting long-term demand can be impossible under some conditions. If you print only enough copies to fill the short-term demand, you do not have to worry about possible changes in public tastes or affluence.

HOW TO READ THE PRINTER LISTINGS

The following pages provide an alphabetical listing of 350 book printers who can print anywhere from 10 to 10,000 (or more) copies of a book or other bound publication. Each listing includes the following information:

Printer's Company Name   Phone Number(s)
Address   Person to Contact
City State Zip Code   Title of Contact Person

Quantities: The minimum, maximum, and optimum quantities of books they can print. Pay particular attention to their optimum number because they can usually give their best prices at this quantity.

Book Sizes: The sizes they normally produce or can produce most economically.

Bindings: [ ] PB = perfect binding (like most softcover books)
[ ] SS = saddlestitching (stapled like a magazine)
[ ] HC = hardcover or casebinding
[ ] C/SB = comb or spiral binding

[I] = They can do this type of binding in-house.
[O] = They can subcontract this type of binding to outside vendors who they work with regularly.
[X] = They can provide this type of binding but did not specify whether in-house or subcontracted.
[ ] = If blank, they don't offer this type of binding.

Capabilities: They can do other printing jobs which are marked with an X like this: [X] Magazines.

Services: They can provide those extra services which are marked with an X like this: [X] Typesetting.

Terms: These are the terms they would offer to a small book publisher, provided the publisher's credit is good.

Following the terms, we provide other comments regarding the printer's special capabilities or services. Where known, we also comment on their reputation, any awards they might have won, how long they have been in business, and other tidbits that might help you to decide whether they can provide the services you require.

User Comments: "Within these quote marks following this heading we give user comments about those printers who were rated by one or more publishers in our spring 1985 user survey. Read these comments carefully; they reflect actual user experiences."

PRINTER RATINGS — SPRING 1985 USER SURVEY

RATINGS	1	2	3	4	5	6	7	8	9	10	Ave
Speed	-	-	-	-	-	-	-	-	-	-	---
Price	-	-	-	-	-	-	-	-	-	-	---
Dependability	-	-	-	-	-	-	-	-	-	-	---
Service . . .	-	-	-	-	-	-	-	-	-	-	---
Quality . . .	-	-	-	-	-	-	-	-	-	-	---
Overall . . .	-	-	-	-	-	-	-	-	-	-	---

The above chart is included in the listings of 50 printers who were rated by over 70 different publishers who responded to our spring 1985 user survey (each publisher rated an average of 2 printers). The ratings were on a scale from 0 to 10, with 5 being an average rating.

The following printers were rated 3 or more times:

Banta	6.17 (6)	Kingsport Press	4.67 (3)
Bookcrafters	6.69 (16)	Kni Book Manufacturers	5.33 (4)
Braun-Brumfield	7.14 (7)	McNaughton & Gunn	6.75 (12)
Capital City Press	5.75 (4)	Malloy Lithograph	7.33 (3)
Delta Litho	8.25 (4)	Thomson-Shore	7.89 (9)
Edwards Brothers	7.0 (8)	Whitehall	4.75 (4)
Inter-Collegiate	6.33 (3)		

The average rating (weighted for number of responses) for the above printers was 6.66. Hence, any printer getting a rating below 6.66 would be considered a below average printer (within this highly used group), but still above average for the universe of all printers (where the average should be 5.0).

Thomson-Shore got the highest rating (7.89) for those printers rated by 7 or more publishers. Braun-Brumfield got the second highest rating (7.14).

Please note, however, that none of these ratings should be taken too seriously because they are not statistically signifi-cant. Use them only as a guide.

38 WAYS TO SAVE ON THE PRODUCTION OF YOUR BOOKS

```
* * * * * * * * * * * * * * * * * * * * * * * * * * * * * * * *
*  Tip:  Tips on how to save money on the production of your   *
*        books are scattered throughout the listings on the   *
*        next 175 pages.  These tips are highlighted by being *
*        boxed in by astericks just like this tip is.         *
* * * * * * * * * * * * * * * * * * * * * * * * * * * * * * * *
```

ABC Printing
13318 NE 12th Avenue
Vancouver WA 98685

206-573-2161
Sales Representative

They asked to be listed in this Directory but were too late to get a full listing.

Academy Books
10 Cleveland Avenue
P O Box 757
Rutland VT 05701

800-451-6045
802-773-9194
Robert A Sharp
General Manager

Quantities: Min: 100   Max: 5000   Opt: 5000

Book Sizes: 5 1/2 x 8 1/2; 6 x 9; 8 1/2 x 11; 9 x 12; 11 x 14

Bindings: [I] PB   [I] SS   [I] HC   [O] C/SB

Capabilities: [X] Magazines   [ ] Galley Copies
                 [X] Journals   [ ] Demand Printing
                 [X] Cookbooks   [ ] 4-color Juvenile Books
                 [ ] Yearbooks   [X] Annual Reports/Brochures
                 [X] Catalogs   [X] Other Commercial Printing

Services: [X] Typesetting   [ ] Teletypesetting
             [ ] Design and Pasteup   [ ] Editing
             [X] 4-color Printing   [X] Warehousing/Shipping

Terms: 1/3 with order, 1/3 with proofs, 1/3 on completion.

A division of Sharp Offset Printing, Academy has been printing books since 1946. They also offer a short-run bindery service (50 books minimum). Call their 800 number for quick quotes.

Access Composition Services
110 S 41st Avenue
Phoenix AZ 85009

602-272-7778
Sam Freedman
Account Executive

They have another plant at 2850 Speer Boulevard, Denver CO 80211; phone 303-458-6955.

Quantities: Min: 250   Max: 10,000   Opt: 1500

Book Sizes: 5 1/2 x 8 1/2 and 8 1/2 x 11 only

Bindings: [I] PB   [I] SS   [O] HC   [O] C/SB

Capabilities: [X] Magazines   [X] Galley Copies
                [X] Journals   [X] Demand Printing
                [X] Cookbooks   [X] 4-color Juvenile Books
                [ ] Yearbooks   [X] Annual Reports/Brochures
                [X] Catalogs   [X] Other Commercial Printing

Services: [X] Typesetting   [X] Teletypesetting
            [ ] Design and Pasteup   [ ] Editing
            [X] 4-color Printing   [X] Warehousing/Shipping

Terms: Net 30 days.

Accurate Web   516-667-3200
355 Marcus Boulevard   Jeri Lynn Selvin Rank
Deer Park   NY 11729   General Manager

Quantities: Min: 3000   Max: none   Opt: not stated

Book Sizes: 5 1/2 x 8 1/2; 6 x 9; 7 1/8 x 9; 8 1/2 x 11

Bindings: [X] PB   [X] SS   [ ] HC   [ ] C/SB

Capabilities: [ ] Magazines   [ ] Galley Copies
                [ ] Journals   [X] Demand Printing
                [X] Cookbooks   [X] 4-color Juvenile Books
                [ ] Yearbooks   [ ] Annual Reports/Brochures
                [X] Catalogs   [X] Other Commercial Printing

Services: [ ] Typesetting   [ ] Teletypesetting
            [ ] Design and Pasteup   [ ] Editing
            [X] 4-color Printing   [X] Warehousing/Shipping

Terms: Not stated.

Ad Infinitum Press   914-664-5930
7 N MacQuesten Parkway   800-942-1992 (New York only)
P O Box 2212   William Brandon,
Mount Vernon   NY 10551   President

    Their 800 number (800-431-1320) is no longer in service.   It
is possible that they have gone out of business. They have never
answered any of our RFQ's or survey forms.

Adams & Abbott Inc.   617-542-1621
46 Summer Street   Robert Rivera
Boston MA 02110   Sales Manager

Quantities: Min: 100   Max: 10,000   Opt: not stated

Book Sizes: 5 x 7; 5 1/2 x 8 1/2; 8 1/2 x 11

Bindings: [X] PB   [X] SS   [ ] HC   [X] C/SB

Capabilities: unknown

Other services: unknown

   They were listed in the last edition of this Directory but did
not answer our current survey. Their speciality is typesetting
and printing foreign language and mathematics books.

The Adams Group   212-255-4900
225 Varick Street   Jim Penders
New York NY 10014   Sales Manager

   They have never answered any of our RFQ's or survey forms.
They apparently do not do short-run book printing.

Adams Press   312-676-3426   SP
30 W Washington Street   Beverly Freid
Chicago IL 60602   Sales Manager

Quantities: Min: 250   Max: 5000   Opt: 1000

Book Sizes: 5 1/2 x 8 1/2; 6 x 9; 8 1/2 x 11; 9 x 12

Bindings: [X] PB   [X] SS   [X] HC   [X] C/SB

Capabilities: [X] Magazines   [ ] Galley Copies
               [ ] Journals   [ ] Demand Printing
               [X] Cookbooks   [ ] 4-color Juvenile Books
               [ ] Yearbooks   [ ] Annual Reports/Brochures
               [X] Catalogs   [ ] Other Commercial Printing

Services: [X] Typesetting   [ ] Teletypesetting
           [ ] Design and Pasteup [ ] Editing
           [X] 4-color Printing   [ ] Warehousing/Shipping

Other services: Will obtain copyright for author and LC numbers.

Terms: Typesetting: 1/2 with order, 1/2 with proofs.
        Printing: Pay in full at time of order.

Adams Press specializes in printing books by self-publishers. In business since 1942, they have advertised regularly in Writer's Digest magazine. They are a reputable and reliable company which is accustomed to working with writers through the mails. They issue a free brochure and price list.

Albany Press   415-428-1800
1343 Powell Street   Sales Representative
Emeryville CA 94608

They did not answer our survey form, but apparently do print short-run books.

Algen Press Corporation   212-463-4605
18-06 130th Street   Lewis Falce
College Point NY 11356   President

They were listed in the last edition of this Directory but did not answer our current survey. They specialize in printing color case covers and dust jackets and may no longer be interested in doing short-run book printing.

Alpine Press   800-343-5901 / 617-341-1800
100 Alpine Circle   Stephen Snyder
Stoughton MA 02072   President

They have never answered any of our RFQ's or survey forms. They apparently do not do short runs.

```
* * * * * * * * * * * * * * * * * * * * * * * * * * * * * * * *
*  TIP:  Don't wait until the last minute.  Rush jobs only    *
*        cause headaches, result in errors of omission, and   *
*        can cost more money in overtime pay and shipping.    *
* * * * * * * * * * * * * * * * * * * * * * * * * * * * * * * *
```

American Lithocraft Corp.   201-434-6617 / 212-344-0177
34 Exchange Place   Carl Yucht
Jersey City NJ 07302   Vice President

Quantities: Min: 1000   Max: 300,000   Opt: 50,000

Book Sizes: any size book.

Bindings: [X] PB   [X] SS   [X] HC   [X] C/SB

Capabilities: [X] Magazines   [ ] Galley Copies
                 [X] Journals   [ ] Demand Printing
                 [X] Cookbooks   [X] 4-color Juvenile Books
                 [ ] Yearbooks   [X] Annual Reports/Brochures
                 [X] Catalogs   [X] Other Commercial Printing

Services: [X] Typesetting   [ ] Teletypesetting
             [X] Design and Pasteup   [ ] Editing
             [X] 4-color Printing   [X] Warehousing/Shipping

Terms: Net 30 days.

A general commercial printer, American Lithocraft emphasizes "high quality and high service." Note that they offer their best prices on quantities of 50,000 or more.

American Offset Printers   213-231-4133
3600 S Hill Street   Attn: Sales Representative
Los Angeles CA 90007

They have never answered any of our RFQ's or survey forms, but according to one of our correspondents, "They are very much alive and well. They do satisfactory work on time (5000 trade paperbacks in 14 days) and are reasonable when problems arise. Their main drawback is that they can only offer good rates using a very limited selection of paper."

American Pizzi Offset Corp.   212-986-1658
141 E 44th Street   Sandro Diani
New York NY 10017   General Manager

They may be brokers for an Italian printer. Since they did not answer our printer survey, we cannot be sure. They probably do not do short runs.

24

Andover Press
516 W 34th Street
New York NY 10001

212-594-3556
Martin Littlefield
Vice President

Quantities: Min: 500   Max: 10,000   Opt: 1000 - 10,000

Book Sizes: 5 1/2 x 8 1/2; 6 x 9

Bindings: [X] PB   [X] SS   [X] HC   [X] C/SB

Capabilities: [ ] Magazines   [ ] Galley Copies
               [ ] Journals   [ ] Demand Printing
               [X] Cookbooks   [X] 4-color Juvenile Books
               [ ] Yearbooks   [ ] Annual Reports/Brochures
               [ ] Catalogs   [ ] Other Commercial Printing

Services: [X] Typesetting   [ ] Teletypesetting
           [X] Design and Pasteup [X] Editing
           [ ] 4-color Printing   [ ] Warehousing/Shipping

Terms: Not stated.

   Andover is the printing arm of Vantage Press, a vanity press.
For more about dealing with vanity presses, send for the free
reprint "Does It Pay to Pay to Have it Published?" from Writer's
Digest, 9933 Alliance Road, Cincinnati, OH 45242. Please include
a self-address, stamped envelope.

==

Anthoensen Press
37 Exchange Street
Portland ME 04112

207-774-3301
Harry N. Milliken
General Manager

   They can print books, periodicals, and catalogs by either
offset or letterpress -- but apparently only in longer runs (they
did not answer our printer survey, so we cannot be sure).

==

AOS Publishing Services
222 West Adams Street
Chicago IL 60606

312-782-6722
Attn: Sales Manager

   They have been advertising their fine design, typography and
printing in Small Press magazine; however, they did not answer
our RFQ or printer survey. Perhaps they are only printing
brokers.

Apollo Books　　　　　　　800-328-8963　　　　　　SP
107 Lafayette　　　　　　　　Robert Kimbril
Winona MN 55987　　　　　　 Vice President Sales

In Minnesota call 800-642-5125, or 507-454-7587.

Quantities: Min: 250　　Max: 50,000　　Opt: 5000

Book Sizes: 4 x 7; 5 1/2 x 8 1/2; 6 x 9; 8 1/2 x 11

Bindings: [I] PB　　[I] SS　　[I] HC　　[I] C/SB

Capabilities:　[X] Magazines　　[X] Galley Copies
　　　　　　　　[X] Journals　　　[ ] Demand Printing
　　　　　　　　[X] Cookbooks　　[X] 4-color Juvenile Books
　　　　　　　　[ ] Yearbooks　　 [X] Annual Reports/Brochures
　　　　　　　　[X] Catalogs　　　[X] Other Commercial Printing

Services:　[X] Typesetting　　　　[X] Teletypesetting
　　　　　　[X] Design and Pasteup　[X] Editing
　　　　　　[X] 4-color Printing　　[X] Warehousing/Shipping

Terms: Net 30 to publishers of 3 or more books annually;
　　　　 otherwise, payment in advance.

In the past, Apollo has been one of the most unreliable book printers in the business. They were continually late on deliveries, repeatedly failed to return phone calls regarding delays, and packed books without adequate protection (so 5 - 10% of the books were damaged during shipment).

They are now under new management and have hired a full-time quality control person and a customer representative. They have also expanded their facilities. It is quite possible that the problems mentioned in the above paragraph were the result of their rapid expansion (as they grew from 6 employees to over 35), but we would still recommend caution in dealing with them since we have just recently received another complaint regarding their work.

Note: They now offer a 2% per day cash rebate for every day they don't meet a deadline (provided you have fulfilled your side of the bargain by returning proofs on time).

User Comments: "I only wish I'd had it (this <u>Directory</u>) before I went to Apollo Books for our last publication. I would have save myself four-plus of the most frustrating months I have ever spent." ... "Very slow. Some pages and covers were off-center." ... "We thought we'd take a chance with them, even after reading your book. The books came over 2 months late. The cover was supposed to be gray. It came out white." ... "They sent my books loose and out of 2000 copies, 235 were damaged."

RATINGS	1	2	3	4	5	6	7	8	9	10	Ave	
Speed	2	-	-	-	-	-	-	-	-	-	---	2
Price	-	-	-	-	-	-	1	-	1	-	---	
Dependability	1	-	1	-	-	-	-	-	-	-	---	
Service . . .	1	-	1	-	-	-	-	-	-	-	---	
Quality . . .	-	1	-	-	1	-	-	-	-	-	---	
Overall . . .	1	-	1	-	-	-	-	-	-	-	---	

Arcata Graphics
101 Merritt / P O Box 6030
Norwalk CT 06852

800-722-7020 / 203-846-6000
Attn: Sales Representative

Arcata Graphics is the second largest printer in the United States with sales of $500,000,000 per year. They have three book printing plants which are capable of doing short runs: Fairfield Graphics, Halliday Lithograph, and Kingsport Press (see the individual listings). They also have other plants that print editions of Reader's Digest, Newsweek, other weekly and monthly magazines, and catalogs.

They have sales offices all over the country:

617-328-3700	Boston MA	305-857-2142	Orlando FL
716-686-2500	Buffalo NY	215-667-8580	Philadelphia PA
312-640-8644	Chicago IL	619-235-4404	San Diego CA
214-942-6168	Dallas TX	415-571-5555	San Francisco CA
213-907-7600	Los Angeles CA	201-233-4148	Westfield NJ
615-385-0460	Nashville TN	301-731-5747	Landover MD
212-840-6468	New York NY	01-242-6214	London England

User's Comments: "Generally good prices on our big books (500+ pages), and good quality. Have had some communications problems with them at times." ... "Gave me the runaround. Delivered months late."

Be sure to also see the comments and ratings under the individual printing plants: Fairfield Graphics, Halliday Lithograph, and Kingsport Press

RATINGS	1	2	3	4	5	6	7	8	9	10	Ave	
Speed	1	-	-	-	1	-	-	-	-	-	---	2
Price	-	-	-	-	-	-	-	1	-	1	---	
Dependability	1	-	-	1	-	-	-	-	-	-	---	
Service . . .	1	-	-	-	1	-	-	-	-	-	---	
Quality . . .	-	-	-	-	-	-	-	2	-	-	---	
Overall . . .	1	-	-	-	-	-	1	-	-	-	---	

D. Armstrong Company
2000 B Governors Circle
Houston TX 77092

800-231-6441
Don Armstrong
President

In Texas call: 800-392-4311, or 713-688-1441.

They were listed in the last edition of this Directory but did not answer our current survey. Their specialty is their typesetting service. Write or call for their Teletext user manual and type selection book. They will also send you a brochure, Book, Budgets, and Authors, which is an excellent summary of the best ways to get the most from your book printer.

Automated Graphic Systems
P O 188 / DeMarr Road
White Plains MD 20695

301-843-1800
Mark Edgar
Sales Marketing Mgr.

Quantities: Min: 500   Max: 50,000   Opt: 20,000

Book Sizes: 5 1/2 x 8 1/2; 6 x 9; 7 x 10; 6 3/4 x 10; 8 1/2 x 11

Bindings: [I] PB   [I] SS   [O] HC   [I] C/SB

Capabilities: [ ] Magazines   [ ] Galley Copies
                [X] Journals   [ ] Demand Printing
                [X] Cookbooks   [X] 4-color Juvenile Books
                [ ] Yearbooks   [ ] Annual Reports/Brochures
                [X] Catalogs   [X] Other Commercial Printing

Services: [X] Typesetting   [X] Teletypesetting
            [X] Design and Pasteup [ ] Editing
            [ ] 4-color Printing   [X] Warehousing/Shipping

Terms: Net 30 days.

They offer total service with art, composition, printing, binding, storage, fulfillment, mailing, and data base management.

B. C. Graphics
176-3 Central Avenue
Farmingdale NY 11735

516-293-9136
Mark Weaver

They asked to be listed in the Directory but were too late to get a full listing.

Banta Company
Curtis Reed Plaza
Menasha WI 54952

414-722-7771
R. J. Cornell
Director of General Books

They have sales offices in the following cities:

404-457-7003	Atlanta GA	714-833-3313	Irvine, CA
617-864-0650	Boston MA	415-966-1800	Los Altos CA
312-297-8800	Chicago IL	212-867-2990	New York NY
216-526-9655	Cleveland OH	703-684-0044	Washington DC

Banta also has a second plant at 3330 Willow Spring Road, Harrisonburg, VA 22801; phone: 703-433-2571.

Quantities: Min: 2500   Max: none   Opt: 2500 - 40,000

Book Sizes: Any size between 4 1/4 x 7 and 8 1/2 x 11.

Bindings: [I] PB   [I] SS   [I] HC   [I] C/SB

Capabilities: [ ] Magazines   [ ] Galley Copies
               [ ] Journals   [ ] Demand Printing
               [ ] Cookbooks   [ ] 4-color Juvenile Books
               [ ] Yearbooks   [ ] Annual Reports/Brochures
               [X] Catalogs   [ ] Other Commercial Printing

Services: [ ] Typesetting   [ ] Teletypesetting
           [ ] Design and Pasteup   [ ] Editing
           [X] 4-color Printing   [X] Warehousing/Shipping

Other services: learning packages including audio/visual
                 materials.

Terms: To be determined.

Banta is a well-regarded printer who does many books for major publishers as well as the smaller independents. They are the twelfth largest printer in the United States, with annual sales of over $300,000,000.

They offer their best prices on runs of 5000 or more, where they can use their Cameron belt press; but, note, this belt press does not reproduce halftones as well as sheet-fed or web presses.

User Comments: "Good average printer, reasonably priced, good service." ... "Good prices on 4 x 7 mass-market paperbacks." ... "Best when quantity is above 5000. Quality not high." ... "Price excellent at quantities over 10,000, but with the rubber belt on their Cameron press half-tones and superfine line art won't look as good as they will with metal plates."

See the results of our ratings survey on the next page.

Banta Company, continued

RATINGS	1	2	3	4	5	6	7	8	9	10	Ave	
Speed	–	–	–	2	2	1	–	1	–	–	5.33	6
Price	–	–	–	–	1	2	–	1	2	–	7.17	
Dependability	–	–	–	–	1	2	–	1	2	–	7.17	
Service . . .	–	–	–	–	2	–	–	2	2	–	7.33	
Quality . . .	–	1	–	–	3	1	1	–	–	–	5.0	
Overall . . .	–	–	–	–	1	3	2	–	–	–	6.17	

Bawden Printing   319-285-4800
400 S 15th Avenue   Howard Stevens
Eldridge IA 52748   Vice President Sales

They have sales offices in Chicago (312-655-5955), Memphis (901-685-1646), and New York (212-302-0009).

Quantities: Min: 300   Max: 500,000   Opt: 20,000

Book Sizes: 5 1/2 x 8 1/2; 6 x 9; 8 1/2 x 11

Bindings: [I] PB   [I] SS   [O] HC   [I] C/SB

Capabilities: [X] Magazines   [ ] Galley Copies
                [X] Journals   [ ] Demand Printing
                [X] Cookbooks   [ ] 4-color Juvenile Books
                [ ] Yearbooks   [X] Annual Reports/Brochures
                [X] Catalogs   [X] Other Commercial Printing

Services: [X] Typesetting   [X] Teletypesetting
            [X] Design and Pasteup   [ ] Editing
            [X] 4-color Printing   [X] Warehousing/Shipping

Terms: Net 30 with approved credit.

Their specialty is "high-quality, short-run book printing with competitive prices and very fast turnaround (e.g., 500 copies of a 300-page book in 4 to 5 days)." They also print tests, such as the ACT tests.

```
* * * * * * * * * * * * * * * * * * * * * * * * * * * * * * *
*  Tip:  Present clean typewritten copy to the typesetter with  *
*        as few editorial changes as possible.  Clean copy      *
*        allows the typesetter to operate faster, resulting in   *
*        a lower charge for the typesetter's time.              *
* * * * * * * * * * * * * * * * * * * * * * * * * * * * * * *
```

Bay Port Press
100 W 35th Street
National City CA 92050

619-420-6296
John W. Collin
President

Quantities: Min: 1000   Max: 20,000   Opt: 10,000

Book Sizes: 5 1/2 x 8 1/2; 8 1/2 x 11

Bindings: [O] PB   [I] SS   [O] HC   [O] C/SB

Capabilities: [X] Magazines   [ ] Galley Copies
               [ ] Journals   [ ] Demand Printing
               [ ] Cookbooks   [ ] 4-color Juvenile Books
               [ ] Yearbooks   [X] Annual Reports/Brochures
               [X] Catalogs   [X] Other Commercial Printing

Services: [X] Typesetting   [ ] Teletypesetting
           [X] Design and Pasteup [ ] Editing
           [X] 4-color Printing   [X] Warehousing/Shipping

Terms: 50% down, balance on delivery.

Bay Port is a full service commercial printer who can print books, catalogs, directories, and manuals in one to four colors. Note that they only do saddlestitching in-house; all other bindings are subcontracted with outside vendors.

W. R. Bean & Son
4800 Frederick Drive NW
Atlanta GA 30378

404-691-5020
Gerry Vaughan
President

W. R. Bean does over $25,000,000 in printing every year, but apparently does not do short-runs.

```
* * * * * * * * * * * * * * * * * * * * * * * * * * * * * * * *
*  Tip:  Consider typesetting some of your books with a type-    *
*        writer or word-processor, as this book was typeset.     *
*        If you plan to market the book primarily to a general   *
*        audience through bookstores, then use regular type-     *
*        setting; but if you plan to market primarily by mail    *
*        or to a specific market (where information is more      *
*        important than looks), you should consider using more   *
*        informal type.  You can save anywhere from $500 to      *
*        $5000 per book by doing your own typesetting on a       *
*        typewriter or computer printer.  More and more books    *
*        are being published in this style.                      *
* * * * * * * * * * * * * * * * * * * * * * * * * * * * * * * *
```

Ovid Bell Press
1201-05 Bluff St
P O Box 381
Fulton MO 65251

800-835-8919 / 314-642-4177
800-642-4117 (Missouri only)
Ovid H. Bell
President

Quantities: Min: 500   Max: 25,000   Opt: 10,000

Book Sizes: 5 1/2 x 8 1/2; 6 x 9; 8 1/2 x 11; and variants.

Bindings: [I] PB   [I] SS   [O] HC   [O] C/SB

Capabilities: [X] Magazines   [ ] Galley Copies
               [X] Journals   [ ] Demand Printing
               [ ] Cookbooks   [ ] 4-color Juvenile Books
               [ ] Yearbooks   [X] Annual Reports/Brochures
               [X] Catalogs   [ ] Other Commercial Printing

Services: [X] Typesetting   [X] Teletypesetting
           [X] Design and Pasteup   [ ] Editing
           [ ] 4-color Printing   [ ] Warehousing/Shipping

Terms: Net 30 with established credit;
        progress payments can be arranged.

They have been in business since 1927 serving customers coast to coast.

Harold Berliner
224 Main Street
Nevada City CA 95959

916-273-2278
Harold A. Berliner
President

Berliner is apparently a letterpress-only printer who does fine limited editions (250 copies) of books, maps, and posters. They did not answer our RFQ or printer survey.

Best Impressions
5350 Cornell Road
Cincinnati OH 45242

800-242-9800 / 513-489-1414
Cliff Hall
Vice President

Quantities: Min: 100   Max: 25,000   Opt: 1000 - 5000

Book Sizes: 5 1/2 x 8 1/2; 6 x 9; 8 1/2 x 11; 11 x 17

Bindings: [I] PB   [I] SS   [O] HC   [I] C/SB

Capabilities: [X] Magazines   [X] Galley Copies
               [X] Journals   [X] Demand Printing
               [X] Cookbooks   [X] 4-color Juvenile Books
               [ ] Yearbooks   [X] Annual Reports/Brochures
               [X] Catalogs   [X] Other Commercial Printing

Services: [X] Typesetting   [ ] Teletypesetting
            [X] Design and Pasteup   [ ] Editing
            [X] 4-color Printing   [X] Warehousing/Shipping

Other services: computer disc duplication.

Terms: 10% discount cash with order; net 30.

They specialize in hi-tech publishing, providing binders and/or boxes for computer software kits. They will also duplicate floppy discs and assemble the kits.

═══

T. H. Best Printing Company   416-447-729
33 Kern Road   J. Kirby Best
Don Mills, Ontario   Assistant General Manager
M3B 1S9   Canada

Quantities: Min: 100   Max: 250,000   Opt: 3000 - 5000

Book Sizes: 5 1/2 x 8 1/2; 6 x 9; 8 1/2 x 11

Bindings: [I] PB   [I] SS   [I] HC   [I] C/SB

Capabilities: [ ] Magazines   [ ] Galley Copies
               [ ] Journals   [ ] Demand Printing
               [X] Cookbooks   [X] 4-color Juvenile Books
               [X] Yearbooks   [ ] Annual Reports/Brochures
               [X] Catalogs   [ ] Other Commercial Printing

Services: [ ] Typesetting   [ ] Teletypesetting
            [ ] Design and Pasteup   [ ] Editing
            [X] 4-color Printing   [ ] Warehousing/Shipping

Terms: 2% 10 days, net 30 with approved credit.

One of Canada's largest book printers, they have won a number of awards for their quality 4-color work. They are accustomed to producing quality books, such as 4-color coffee-table books, fine scholarly books, and limited editions with padded covers, foil stamping, and other extras. They have printed books for most major Canadian publishers and are willing to print for U.S. publishers as well.

Bireline Publishing Company
220 S Fulton / P O Box 415
Newell IA 50568

712-272-4417
John R. Bireline
President

SP

They have not answered our last two printer surveys, yet they are still in business because we know of at least one person who has used them recently. However, because they have not answered our surveys, we can only give you limited information about their capabilities.

They specialize in print runs of 100 to 1000 copies of self-published books with 24 to 128 pages. Perfect binding costs up to 40 cents extra per book.

User Comments: "Thoroughly honest and quite nice to work with." ... "While he was not the least expensive printer, he was the most responsive. ... I can't say enough kind words about the way my job was handled. I've been kept fully posted, made aware of a small problem (after he took care of resolving it for me; just as I asked him to do if such a problem occurred)."

RATINGS	1	2	3	4	5	6	7	8	9	10	Ave	
Speed	-	-	-	-	1	-	-	-	-	-	---	1
Price	-	-	-	-	-	-	1	-	-	-	---	
Dependability	-	-	-	-	-	-	1	-	-	-	---	
Service . . .	-	-	-	-	-	-	-	-	1	-	---	
Quality . . .	-	-	-	-	-	-	-	1	-	-	---	
Overall . . .	-	-	-	-	-	-	1	-	-	-	---	

Blake Printery
2222 Beebee Street
San Luis Obispo CA 93401

805-543-6843
Richard Blake
President

California customers may call a free in-state 800 number -- 800-792-6946.

Quantities: Min: 1000   Max: open   Opt: 5000

Book Sizes: Almost any size.

Bindings: [O] PB   [I] SS   [O] HC   [I] C/SB

Capabilities: [X] Magazines   [ ] Galley Copies
                [X] Journals   [ ] Demand Printing
                [X] Cookbooks   [ ] 4-color Juvenile Books
                [ ] Yearbooks   [X] Annual Reports/Brochures
                [X] Catalogs   [X] Other Commercial Printing

Services: [X] Typesetting   [ ] Teletypesetting
         [X] Design and Pasteup [X] Editing
         [X] 4-color Printing   [ ] Warehousing/Shipping

Terms: Negotiable.

They print high quality full-color books and other multiple page products, plus postcards and fine art posters.

===

Bolger Publications   612-645-6311
3301 Como Avenue SE   Jack Bolger
Minneapolis MN 55414   President

They did not answer our printer survey, but we are told that they produce high quality (but not cheap) magazines, directories, and price lists.

===

The Book Press   802-257-7701
Putney Road   David Rapisardi
Brattleboro VT 05301   Vice President Sales

Book Press has sales of some $25,000,000 per year. They won a number of awards in 1984 for their binding. However, they did not answer our printer surveys, so we cannot give you any other details concerning their capabilities. It is possible that they are not interested in doing short-runs.

```
* * * * * * * * * * * * * * * * * * * * * * * * * * * * * * *
*  Tip:  If you have a computer or word-processor and want to  *
*        have a professionally typeset book, consider using a  *
*        compositor or printer capable of typesetting direct   *
*        from your disk or via telecommunications (see the     *
*        the classified index following these listings).  You  *
*        can save both time and money since you are, in effect,*
*        doing the typesetting input as you prepare the book's *
*        manuscript with your computer.                        *
*           According to the National Composition Association, *
*        keyboarding the original input and proofreading make  *
*        up 53% of the typical costs of regular typesetting.   *
*        Corrections account for another 13% of typical costs. *
*        By providing your own proofread and corrected input,  *
*        you can save up to 66% of total typesetting costs.    *
*        Check with your printer or typesetter to see how much *
*        you can save by doing your own keyboarding.           *
* * * * * * * * * * * * * * * * * * * * * * * * * * * * * * *
```

Book-Mart Press
2001 42nd Street
North Bergen NJ 07047

201-864-1887 / 212-594-3344
Michelle A. Gluckow
Vice President, Marketing

Quantities: Min: 300   Max: 25,000   Opt: 1500 - 5000

Book Sizes: 5 1/2 x 8 1/2; 6 x 9; 7 x 10; 8 1/2 x 11

Bindings: [I] PB   [I] SS   [O] HC   [I] C/SB

Capabilities: [ ] Magazines   [ ] Galley Copies
               [X] Journals   [ ] Demand Printing
               [ ] Cookbooks   [ ] 4-color Juvenile Books
               [ ] Yearbooks   [ ] Annual Reports/Brochures
               [X] Catalogs   [ ] Other Commercial Printing

Services: [X] Typesetting   [ ] Teletypesetting
            [ ] Design and Pasteup [ ] Editing
            [ ] 4-color Printing   [X] Warehousing/Shipping

Terms: To be arranged.

Book-Mart specializes in the production of softcover journals
and books. They "offer excellent service, quality workmanship,
and competitive pricing." They also promise quick turn-arounds.
We have never heard anything to contradict their assertions; they
seem to be a good company to work with.

═══

BookCrafters
140 Buchanan St / P O Box 370
Chelsea MI 48118

313-475-9145
Kathy King
Marketing Executive

They have sales offices in a number of different areas:

312-298-0557 Chicago, IL   212-557-1220 New York, NY
415-363-8422 California (north)   801-226-6002 Southwest U.S.
714-621-6198 California (south)   703-276-8666 Washington DC

They also have another printing plant at Lee Hill Industrial
Park, P O Box 892, Fredericksburg, VA 22401; phone: 703-371-3800.

Quantities: Min: 250   Max: open   Opt: 500 - 10,000

Book Sizes: Almost any size, including custom sizes.

Bindings: [I] PB   [O] SS   [I] HC   [I] C/SB

For more details, see the page following Book-Mart's ad.

BookCrafters continued

Capabilities: [ ] Magazines   [ ] Galley Copies
                [X] Journals   [X] Demand Printing
                [X] Cookbooks   [X] 4-color Juvenile Books
                [ ] Yearbooks   [ ] Annual Reports/Brochures
                [X] Catalogs   [ ] Other Commercial Printing

Services: [X] Typesetting   [X] Teletypesetting
          [X] Design and Pasteup [ ] Editing
          [X] 4-color Printing   [X] Warehousing/Shipping

Other services: audio and video duplication.

Terms: Subject to qualification.

BookCrafters has an uneven reputation. Some people swear by them; others swear at them. They were highly rated in the 1981 book printers user survey done by <u>Small Press Review</u>, and they were used by more people than any other press in our own survey conducted this year.

It seems that they do superb work for regular and/or knowledgeable publishers, but have problems servicing neophytes and self-publishers. We recommend caution in using them unless you intend to become a regular customer of theirs.

User Comments: "They promise 20 working days delivery, but 90 days is more to the truth. I work with them because they give us extra time to pay." ... "They do screw up occasionally. Nice to work with. Gets prices to me quickly." ... "Our standby -- always dependable and reasonably priced." ... "Slow (40 days). Reasonable price. Great quality." ... "My favorite press. They easily produce the best quality of any of our printers, their prices are very competitive, but the best thing about them is their service. They're the only printer I've heard of that gives you a competitive schedule, and then finishes ahead of it." ... "Printing quality is good, but their sales reps put customers last. $600 difference in our quote and final job!" ... "A very average company with serious quality control problems and an even more serious lack of sales management." ... "They got the book out exactly when promised. Excellent to work with." ... "They did a shoddy job, printed the job through to completion without sending any sort of review material or checking with me if their low-quality result would suffice, then charged me full price for the work done even tho it was thrown away." ... "Seven months from contract signing until books arrived. The so-called bluelines were so poor I had to phone the president who, twice, had to expedite matters when the dragging continued. The final books were excellent." ... "A pleasure to work with. Excellent quality. Competitive price. Accommodating with some last minute changes we made. Good Folk!"

RATINGS	1	2	3	4	5	6	7	8	9	10	Ave	
Speed	1	-	-	2	6	3	3	-	1	-	5.44	16
Price	-	-	-	1	2	2	4	4	2	1	7.13	
Dependability	-	1	2	1	-	3	1	3	2	2	6.19	
Service . . .	-	1	1	-	2	-	2	5	4	1	7.19	
Quality . . .	-	-	1	1	2	-	1	6	3	2	7.44	
Overall . . .	-	-	1	3	1	-	3	6	2	-	6.69	

BookMasters
638 Jefferson St / P O Box 159
Ashland OH 44805

800-537-6727 / 419-289-6051
Lynn Smith
Marketing

BookMasters are printing brokers with complete typesetting and graphic facilities including teletypesetting and optical scanning equipment. They are national representatives of BookCrafters (see BookCrafters for details on their capabilities and services). Call their 800 number for quick quotes.

William Boyd Printing Company
49 Sheridan Avenue
Albany NY 12210

518-436-9686 / 212-757-5530
Carl Johnson
Vice President

Quantities: Min: 1000   Max: 100,000   Opt: not stated

Book Sizes: 5 1/2 x 8 1/2; 6 x 9; 7 x 10; 8 1/2 x 11

Bindings: [I] PB   [I] SS   [X] HC   [X] C/SB

Capabilities: [X] Magazines   [ ] Galley Copies
                [X] Journals   [ ] Demand Printing
                [X] Cookbooks   [X] 4-color Juvenile Books
                [X] Yearbooks   [ ] Annual Reports/Brochures
                [X] Catalogs   [X] Other Commercial Printing

Services: [X] Typesetting   [ ] Teletypesetting
            [X] Design and Pasteup   [ ] Editing
            [X] 4-color Printing   [X] Warehousing/Shipping

Terms: Net 30 with approved credit.

Founded in 1889, they print United Nations publications, lawbooks, scholarly journals, and commercial work as well as trade paperbacks.

Braceland Brothers
7625 Suffolk Avenue
Philadelphia PA 19153

215-492-0200
B. T. Sweeney
National Sales Manager

Braceland is a commercial printer with annual sales of over $25,000,000. They also print books and catalogs, and can provide composition and fulfillment services. However, they have never answered any of our RFQ's or printer surveys; hence, either they do not do short-runs or their sales department needs to revamp their inquiry processing systems.

User Comment: "Extremely kind and courteous to work with."

RATINGS	1	2	3	4	5	6	7	8	9	10	Ave	
Speed	-	-	-	-	-	-	1	-	-	-	---	1
Price	-	-	-	-	-	-	-	-	-	1	---	
Dependability	-	-	-	-	-	-	-	-	-	1	---	
Service . . .	-	-	-	-	-	-	-	-	-	1	---	
Quality . . .	-	-	-	-	-	-	1	-	-	-	---	
Overall . . .	-	-	-	-	-	-	-	1	-	-	---	

Braun-Brumfield Inc.
100 N Staebler Rd / P O 1203
Ann Arbor MI 48106

313-662-3291
J. E. Cooch
Customer Service Manager

A division of Heritage Communications, they have sales representatives in Arlington VA (703-522-5582), Chicago, New York, San Francisco CA (415-974-5030), and Yardley/Philadelphia PA (215-493-8849).

Quantities: Min: 100   Max: 20,000   Opt: 500 - 2500

Book Sizes: Almost any size between 5 x 7 and 9 x 12.

Bindings: [I] PB   [I] SS   [I] HC   [I] C/SB

Capabilities: [ ] Magazines   [ ] Galley Copies
               [X] Journals   [ ] Demand Printing
               [ ] Cookbooks   [ ] 4-color Juvenile Books
               [ ] Yearbooks   [ ] Annual Reports/Brochures
               [ ] Catalogs   [ ] Other Commercial Printing

Services: [X] Typesetting   [ ] Teletypesetting
           [ ] Design and Pasteup   [ ] Editing
           [ ] 4-color Printing   [X] Warehousing/Shipping

Terms: Net 30 with approved credit.

Braun-Brumfield is noted for its high quality book production. They were rated among the top ten book printers in the 1981 _Small Press Review_ survey and came in second with an overall average of 7.14 on our user's survey this year.

Send for their excellent resource booklets: _Book Manufacturing Glossary_, _Book Paper Samples_, and _Type Sample Book_.

User Comments: "High quality, good scheduling, helpful service. Somewhat limited paper selection. Eager to please and very helpful." ... "The most expensive; quality high." ... "Good quality." ... "Good quality work, but missed shipping date by 2 weeks without notifying us."

RATINGS	1	2	3	4	5	6	7	8	9	10	Ave	
Speed	-	-	-	1	4	-	-	2	-	-	5.71	7
Price	-	1	-	-	1	1	-	2	2	-	6.71	
Dependability	-	-	-	1	1	-	2	2	-	1	7.00	
Service . . .	-	-	-	1	3	-	-	1	1	1	6.57	
Quality . . .	-	-	-	1	-	-	2	2	-	2	7.71	
Overall . . .	-	-	-	-	2	1	-	3	-	1	7.14	

Brennan Printing　　　　　　515-595-2000
100 Main Street　　　　　　　　Robert Brennan
Deep River IA 52222　　　　　 Owner

Quantities: Min: none   Max: 20,000   Opt: 5000 - 10,000

Book Sizes: 5 1/2 x 8 1/2; 6 x 9; 8 1/2 x 11

Bindings: [X] PB   [X] SS   [ ] HC   [I] C/SB

Capabilities: [X] Magazines   [ ] Galley Copies
　　　　　　　 [ ] Journals   [ ] Demand Printing
　　　　　　　 [X] Cookbooks   [ ] 4-color Juvenile Books
　　　　　　　 [ ] Yearbooks   [X] Annual Reports/Brochures
　　　　　　　 [X] Catalogs   [X] Other Commercial Printing

Services: [X] Typesetting   [ ] Teletypesetting
　　　　　 [ ] Design and Pasteup [ ] Editing
　　　　　 [ ] 4-color Printing   [X] Warehousing/Shipping

Terms: 25% on receipt of copy; balance net 30.

Comb-bound cookbooks are their specialty. Their work is good but their prices for perfectbound books are higher than most. Check them out, though, if you're doing a comb-bound book.

R. L. Bryan Company
301 Greystone Blvd / P O 368
Columbia SC 29202

803-779-3560
Jack Whitesides
Vice President Sales

Quantities: Min: 500   Max: 25,000   Opt: 5000 - 10,000

Book Sizes: 5 1/2 x 8 1/2; 6 x 9; 7 x 10; 8 1/2 x 11; 9 x 12

Bindings: [I] PB   [I] SS   [O] HC   [I] C/SB

Capabilities: [X] Magazines   [ ] Galley Copies
               [X] Journals   [ ] Demand Printing
               [ ] Cookbooks   [ ] 4-color Juvenile Books
               [ ] Yearbooks   [X] Annual Reports/Brochures
               [X] Catalogs   [X] Other Commercial Printing

Services: [X] Typesetting   [X] Teletypesetting
           [X] Design and Pasteup [ ] Editing
           [X] 4-color Printing   [X] Warehousing/Shipping

Terms: Net 30 with approved credit.

William Byrd Press
2901 Byrdhill Road / P O 27481
Richmond VA 23261

804-264-2711
Dave Wilson
Vice President Sales

They have sales offices in Hartford CT (203-724-1973), New York NY (212-557-1505), Old Bridge NJ (201-591-1380), Springfield VA (703-321-8610), and Washington DC (202-833-1054).

They are one of the top fifty printers in the United States, with sales of over $40,000,000 per year. They must, however, specialize in longer runs because they have never answered any of our RFQ's or printer surveys.

C & M Press
850 East 73rd Ave #12
Thornton CO 80229

303-289-4757
Dean Carroll
Manager

Quantities: Min: 50   Max: 10,000   Opt: 500

Book Sizes: 5 1/2 x 8 1/2; 6 x 9; 8 1/2 x 11; 11 x 17

Bindings: [I] PB   [I] SS   [I] HC   [I] C/SB

Capabilities: [X] Magazines   [ ] Galley Copies
               [X] Journals   [X] Demand Printing
               [ ] Cookbooks   [ ] 4-color Juvenile Books
               [ ] Yearbooks   [X] Annual Reports/Brochures
               [ ] Catalogs   [X] Other Commercial Printing

Services: [ ] Typesetting   [ ] Teletypesetting
           [X] Design and Pasteup   [ ] Editing
           [ ] 4-color Printing   [X] Warehousing/Shipping

Terms: 1/3 down, 1/3 with proofs, 1/3 on delivery.

They specialize in ultra-short runs (500 or less) of computer manuals, reports, theses, club histories, and self-publications.

Caldwell Printers   818-447-4601
29 S First Avenue   Bob E. Caldwell
Arcadia CA 91006   Co-owner

Quantities: Min: 50   Max: 5000   Opt: 2000

Book Sizes: 5 1/2 x 8 1/2; 6 x 9; 8 1/2 x 11; 9 x 12

Bindings: [O] PB   [I] SS   [O] HC   [ ] C/SB

Capabilities: [ ] Magazines   [ ] Galley Copies
               [ ] Journals   [ ] Demand Printing
               [ ] Cookbooks   [ ] 4-color Juvenile Books
               [ ] Yearbooks   [ ] Annual Reports/Brochures
               [X] Catalogs   [X] Other Commercial Printing

Services: [X] Typesetting   [ ] Teletypesetting
           [X] Design and Pasteup   [ ] Editing
           [ ] 4-color Printing   [ ] Warehousing/Shipping

Terms: 50% deposit, balance on delivery.

Caldwell is a family-owned business of "conscientious crafts-men" who print short runs of manuals, poetry and autobiographies.

```
* * * * * * * * * * * * * * * * * * * * * * * * * * * * * * * *
*  Tip:  Seek the help of your printers in cutting costs.      *
*        Talk to them early in the planning of the layout and  *
*        design of your books.  They may be able to suggest    *
*        minor changes in your specs that will save you money  *
*        without affecting the quality of your books.          *
* * * * * * * * * * * * * * * * * * * * * * * * * * * * * * * *
```

Canterbury Press
301 Mill Street
Rome NY 13440

315-337-5900
Alison Snider
Customer Service Manager

Quantities: Min: 500   Max: 25,000   Opt: 5000

Book Sizes: 5 1/2 x 8 1/2; 6 x 9; 7 x 10; 8 1/2 x 11

Bindings: [I] PB   [I] SS   [O] HC   [I] C/SB

Capabilities: [X] Magazines   [ ] Galley Copies
               [X] Journals   [X] Demand Printing
               [X] Cookbooks   [ ] 4-color Juvenile Books
               [ ] Yearbooks   [X] Annual Reports/Brochures
               [X] Catalogs   [X] Other Commercial Printing

Services: [X] Typesetting   [ ] Teletypesetting
           [X] Design and Pasteup [ ] Editing
           [X] 4-color Printing   [X] Warehousing/Shipping

Terms: Net 30.

Canterbury is an employee-owned company which prints booklets, brochures, catalogs, workbooks, handbooks and other publications. They offer design services as well as typesetting, printing, and binding.

Capital City Press
P O Box 546
Montpelier VT 05602

802-223-5207
Glennis Drew
Estimator

Quantities: Min: 500   Max: 25,000   Opt: 5000

Book Sizes: Many sizes between 5 1/2 x 8 1/2 and 9 x 12.

Bindings: [I] PB   [I] SS   [I] HC   [O] C/SB

Capabilities: [X] Magazines   [ ] Galley Copies
               [X] Journals   [ ] Demand Printing
               [X] Cookbooks   [X] 4-color Juvenile Books
               [X] Yearbooks   [ ] Annual Reports/Brochures
               [X] Catalogs   [ ] Other Commercial Printing

Services: [X] Typesetting   [X] Teletypesetting
           [ ] Design and Pasteup [ ] Editing
           [ ] 4-color Printing   [X] Warehousing/Shipping

Terms: 2% 10, net 30 days with approved credit.

Capital City has been in business since 1908 and currently does about $10,000,000 worth of business a year. They print over 140 journals (medical, technical, and literary) as well as trade paperbacks and hardcovers. They do excellent work and offer very reasonable prices (they gave us one of the lowest quotes for the printing of the Second Edition of this <u>Directory</u>). Their repro work on photos and illustrations is superb.

They itemize their quotes so you know all the costs (such specifics as covers, prep work, plates, proofs, press work, binding, paper, cartons, and shipping). Hence, if you decide to change one of the specifications later on, you can better estimate what that change will cost you.

User Comments: "They have had some problems. Dropped them for awhile. I'm trying them again. Slowly." ... "They promise 20 working days, but 90 is more to the truth." ... "Excellent on difficult jobs. A little more expensive; they make up for it in speed and customer service." ... "The responsiveness and very believable (and demonstrated) concern by Capital City to maintain their good name and customer goodwill impressed me greatly. They were willing to do anything we asked to correct their earlier mistakes. I have never done business with a more convincingly concerned company!"

RATINGS	1	2	3	4	5	6	7	8	9	10	Ave	
Speed	1	-	-	1	-	-	-	2	-	-	5.25	4
Price	-	-	-	-	1	-	-	2	1	-	7.5	
Dependability	-	1	1	-	-	-	1	-	-	1	5.5	
Service . . .	-	1	-	-	-	-	1	1	1	-	6.5	
Quality . . .	-	-	-	-	1	-	-	1	2	-	7.75	
Overall . . .	-	-	-	1	1	-	2	-	-	-	5.75	

Carnes Publications Services
23811 Chagrin Blvd #LL64
Beachwood OH 44122

216-292-7959
William L. Carnes
President

They are printing brokers specializing in technical and professional publications. They have not answered our last two printer surveys and may no longer be in business.

```
* * * * * * * * * * * * * * * * * * * * * * * * * * * * * * *
*  Tip:  Prepare your camera-ready copy so all the pages can   *
*        be shot using the same camera setting (requiring no   *
*        no special reductions or enlargements).               *
* * * * * * * * * * * * * * * * * * * * * * * * * * * * * * *
```

Case-Hoyt Corporation
800 St Paul Street
Rochester NY 14601

716-232-6840
Robert Frame
Vice President Sales

Case-Hoyt, with an annual sales volume of $90,000,000, is among the 25 largest printers in the United States. They specialize in printing 4-color catalogs and other publications. However, they do not do press runs of less than 10,000 copies.

CBP Press
2700 Pine Street
St Louis MO 63166

314-371-6900
Donald Piere
Sales Manager

Quantities: Min: 300   Max: 10,000   Opt: 5000 - 7000

Book Sizes: 5 1/2 x 8 1/2; 6 x 9; 7 x 10; 8 1/2 x 11

Bindings: [I] PB   [I] SS   [O] HC   [I] C/SB

Capabilities: [X] Magazines   [ ] Galley Copies
                [X] Journals   [ ] Demand Printing
                [X] Cookbooks   [ ] 4-color Juvenile Books
                [ ] Yearbooks   [X] Annual Reports/Brochures
                [X] Catalogs   [X] Other Commercial Printing

Services: [ ] Typesetting   [ ] Teletypesetting
           [ ] Design and Pasteup [ ] Editing
           [X] 4-color Printing   [X] Warehousing/Shipping

Terms: 1/3 down, 1/3 with proofs, 1/3 on delivery.

CBP used to print lots of books, but now they are doing more general commercial printing: posters, magazines, brochures -- and some books (though it is no longer their specialty).

Central Publishing Company
401 N College Avenue
Indianapolis IN 46202

317-636-4504
Vona W. Lauman
Vice President

Quantities: Min: 250   Max: 20,000   Opt: 10,000

Book Sizes: 5 1/2 x 8 1/2; 6 x 9; 8 1/2 x 11

Bindings: [I] PB   [I] SS   [O] HC   [I] C/SB

Capabilities: [X] Magazines   [X] Galley Copies
               [X] Journals   [ ] Demand Printing
               [X] Cookbooks   [X] 4-color Juvenile Books
               [ ] Yearbooks   [X] Annual Reports/Brochures
               [X] Catalogs   [X] Other Commercial Printing

Services: [X] Typesetting   [X] Teletypesetting
           [X] Design and Pasteup [ ] Editing
           [X] 4-color Printing   [X] Warehousing/Shipping

Terms: Discuss on an individual basis.

They have been printers for over 60 years.

Coach House Press   416-919-2217
401 Huron St (Rear)   Stan Bevington
Toronto, Ontario   President
M5S 2G5   Canada

Quantities: Min: 500   Max: 5000   Opt: 1000

Book Sizes: 5 1/2 x 8 1/2 only

Bindings: [X] PB   [ ] SS   [X] HC   [ ] C/SB

Capabilities: [ ] Magazines   [ ] Galley Copies
               [X] Journals   [ ] Demand Printing
               [ ] Cookbooks   [ ] 4-color Juvenile Books
               [ ] Yearbooks   [ ] Annual Reports/Brochures
               [ ] Catalogs   [ ] Other Commercial Printing

Services: [X] Typesetting   [X] Teletypesetting
           [ ] Design and Pasteup [ ] Editing
           [X] 4-color Printing   [ ] Warehousing/Shipping

Terms: Net.

Coach House is a small Canadian publisher who also does printing for other small presses. See article about them in Small
Press, January 1985 (pp. 45-48).

```
* * * * * * * * * * * * * * * * * * * * * * * * * * * * * * * *
*  Tip:  Try to make all your editorial changes before you    *
*        send your copy to the typesetter.  Making changes     *
*        after the copy has been typeset will cost you at      *
*        least two to three times as much.                     *
* * * * * * * * * * * * * * * * * * * * * * * * * * * * * * * *
```

Coleman Graphics
99 Milbar Boulevard
Farmingdale NY 11735

516-293-0383
Saul Steinberg
Director

   Coleman is a publisher of holistic literature who will also do printing for other holistic or new age publishers. They did not answer our printer survey forms, so we cannot give you any further details regarding their capabilities and services.

Colortone Press
2400 17th Street NW
Washington DC 20009

202-387-6800
Al J. Hackl
President

Quantities: Min: none   Max: none   Opt: not stated

Book Sizes: 5 1/2 x 8 1/2; 6 x 9; 8 x 9; 8 1/2 x 11

Bindings: [X] PB   [X] SS   [X] HC   [X] C/SB

Capabilities: [X] Magazines   [ ] Galley Copies
               [ ] Journals   [ ] Demand Printing
               [X] Cookbooks   [X] 4-color Juvenile Books
               [ ] Yearbooks   [X] Annual Reports/Brochures
               [X] Catalogs   [X] Other Commercial Printing

Services: [X] Typesetting   [X] Teletypesetting
           [X] Design and Pasteup [X] Editing
           [X] 4-color Printing   [ ] Warehousing/Shipping

Terms: Not stated.

   Colortone is associated with Acropolis Press (publisher of Color Me Beautiful and other books). They specialize in printing books requiring high quality work, from photo and art books to any 4-color books. They are not cheap, but they are good.

Columbia Planograph
10126 Bacon Drive
Beltsville MD 20705

301-937-4677 / 212-564-9685
Donald A. Eckert
President

   They are short to long-run printers of magazines and paperback books. However, they have never answered any of our RFQ's or printer surveys, so we cannot be sure how serious they are about doing short runs.

48

Commercial Printing Company   503-773-7575 / 503-228-1182
2661 S Pacific Hwy / P O 1165   Cleve Tooker
Medford OR 97501   Vice President of Operations

Quantities: Min: 500   Max: 100,000   Opt: 20,000

Book Sizes: 4 5/8 x 6 5/8; 5 1/2 x 8 1/2; 6 x 9; 8 1/2 x 7;
             8 1/2 x 11; 9 x 12

Bindings: [I] PB   [I] SS   [I] HC   [I] C/SB

Capabilities: [ ] Magazines   [ ] Galley Copies
                 [X] Journals   [ ] Demand Printing
                 [ ] Cookbooks   [ ] 4-color Juvenile Books
                 [ ] Yearbooks   [X] Annual Reports/Brochures
                 [X] Catalogs   [X] Other Commercial Printing

Services: [X] Typesetting   [X] Teletypesetting
           [X] Design and Pasteup   [X] Editing
           [X] 4-color Printing   [X] Warehousing/Shipping

Terms: Net 30 with approved credit.

Founded in 1906, they have been printing books since 1974.
They specialize in 4-color coffee table books (under 10,000
copies) and computer software manuals. They pride themselves on
being "very flexible."

===

Community Press   801-225-2299
5600 N University Avenue   Ron Baker
Provo UT 84504   Estimator

They also have sales offices in California (415-494-8051),
southwest Oregon (206-577-1041), and Portland, OR (503-253-0063).

Quantities: Min: 500   Max: 50,000   Opt: 10,000

Book Sizes: Almost any size, including all standard sizes.

Bindings: [O] PB   [I] SS   [I] HC   [I] C/SB

Capabilities: [X] Magazines   [ ] Galley Copies
                 [X] Journals   [ ] Demand Printing
                 [X] Cookbooks   [X] 4-color Juvenile Books
                 [X] Yearbooks   [X] Annual Reports/Brochures
                 [X] Catalogs   [X] Other Commercial Printing

See the next page for more about their services.

Community Press continued

Services: [X] Typesetting   [ ] Teletypesetting
            [X] Design and Pasteup   [ ] Editing
            [X] 4-color Printing   [X] Warehousing/Shipping

Terms: Net 30 with approved credit.

   They print newsletters, brochures, and annual reports as well
as books, journals, and yearbooks. Their California salesman
says that their prices are "too good to be true." We'd certainly
recommend that West Coast publishers check it out to verify the
truth of his statement. Be sure to ask for samples of their
work.

Comput-A-Print   702-786-2300
1040 Matley Lane #3   Richard S. Basham Jr.
Reno NV 89502   President

Quantities: Min: 50   Max: 10,000   Opt: 2500

Book Sizes: 5 1/2 x 8 1/2; 6 x 9; 8 1/2 x 11

Bindings: [X] PB   [X] SS   [ ] HC   [X] C/SB

Capabilities: [ ] Magazines   [X] Galley Copies
                [X] Journals   [X] Demand Printing
                [ ] Cookbooks   [ ] 4-color Juvenile Books
                [ ] Yearbooks   [X] Annual Reports/Brochures
                [ ] Catalogs   [X] Other Commercial Printing

Services: [X] Typesetting   [ ] Teletypesetting
            [X] Design and Pasteup   [ ] Editing
            [ ] 4-color Printing   [ ] Warehousing/Shipping

Terms: 50% down, balance C.O.D.; or 5% 10, net 30 with credit
        approval.

Concepts Unlimited   617-263-6777
P O Box 111   Elaine Bonneau
Acton MA 01720   President

   They apparently print 4-color magazines, catalogs, and books
in short-runs, but they did not answer our RFQ or printer survey.

Coneco Laser Graphics
58 Dix Avenue
Glen Falls NY 12801

518-793-3823
Carman P. Elliott
Production Editor

Quantities: Min: 10   Max: 2500   Opt: 500

Book Sizes: Any size up to 8-1/2x11.

Bindings: [I] PB   [I] SS   [O] HC   [O] C/SB

Capabilities: [ ] Magazines   [X] Galley Copies
               [X] Journals   [X] Demand Printing
               [X] Cookbooks   [ ] 4-color Juvenile Books
               [ ] Yearbooks   [X] Annual Reports/Brochures
               [ ] Catalogs   [ ] Other Commercial Printing

Services: [X] Typesetting   [X] Teletypesetting
           [ ] Design and Pasteup [ ] Editing
           [ ] 4-color Printing   [ ] Warehousing/Shipping

Terms: Not stated.

They can typeset from manuscript, scanner, disk, or modem.
They are ultra-short-run specialists. Try them if you only need
a few copies (less than 1000) of your book or other publication,
or if you need fast delivery. They can print and bind galley
copies within 14 days.

Connecticut Printers
55 Granby Street
Bloomfield CT 06002

203-242-0711
Ken Lima
Vice President Sales

They specialize in producing educational and trade catalogs,
but can also print books and magazines (1 to 4 colors, either
perfectbound or saddlestitched). They did not answer our printer
survey, so we cannot give you any other details.

Consolidated Printers
2630 - 8th Street
Berkeley CA 94710

415-843-8524
Attn: Sales Representative

They are listed in the Berkeley phone book as book printers,
but they have not answered our printer surveys. Perhaps they are
not interested in doing short runs.

Contemporary Lithographers
1501 S Blount Street
Raleigh NC 27603

919-821-2211
Jerry Cooper
Sales Manager

They are listed in LMP as being short to medium run book printers, softcover or casebound. However, as a matter of company policy, they asked not to have their capabilities listed in this Directory. They don't like to be estimators for single job publishers (in other words, they do not appreciate receiving shot-gun RFQ's).

User Comment: "I do not recommend."

RATINGS	1	2	3	4	5	6	7	8	9	10	Ave	
Speed	1	-	-	-	-	-	-	-	-	-	---	1
Price	-	-	-	-	-	-	-	1	-	-	---	
Dependability	1	-	-	-	-	-	-	-	-	-	---	
Service . . .	1	-	-	-	-	-	-	-	-	-	---	
Quality . . .	1	-	-	-	-	-	-	-	-	-	---	
Overall . . .	-	1	-	-	-	-	-	-	-	-	---	

Copen Press
100 Berriman Street
Brooklyn NY 11208

212-235-4270
Irwin Isquith
President

Quantities: Min: 5000   Max: 1,000,000   Opt: not stated

Book Sizes: 5 1/2 x 8 1/2; 8 1/2 x 11 (or slightly smaller)

Bindings: [I] PB   [I] SS   [ ] HC   [ ] C/SB

Capabilities: [ ] Magazines   [ ] Galley Copies
                [ ] Journals   [X] Demand Printing
                [ ] Cookbooks   [ ] 4-color Juvenile Books
                [ ] Yearbooks   [ ] Annual Reports/Brochures
                [ ] Catalogs   [X] Other Commercial Printing

Services: [ ] Typesetting   [ ] Teletypesetting
            [ ] Design and Pasteup   [ ] Editing
            [ ] 4-color Printing   [ ] Warehousing/Shipping

Terms: Payment with order.

They publish a standard price list which they will send to you on request. They specialize in printing one or two-color books, booklets, and newsprint catalogs. They can also print tabloid-sized publications.

Corley Printing Company
9804 Page Boulevard
St Louis MO 63132

314-426-3900
David C. Deibel
Vice President

Quantities: Min: 500   Max: 100,000   Opt: 1000 - 25,000

Book Sizes: 5 1/2 x 8 1/2; 6 x 9; 8 1/2 x 11 (their optimum)

Bindings: [I] PB   [I] SS   [ ] HC   [0] C/SB

Capabilities:
- [ ] Magazines
- [ ] Journals
- [ ] Cookbooks
- [ ] Yearbooks
- [X] Catalogs

- [ ] Galley Copies
- [ ] Demand Printing
- [ ] 4-color Juvenile Books
- [ ] Annual Reports/Brochures
- [ ] Other Commercial Printing

Services:
- [ ] Typesetting
- [ ] Design and Pasteup
- [ ] 4-color Printing

- [ ] Teletypesetting
- [ ] Editing
- [ ] Warehousing/Shipping

Terms: 1% 10, net 30 with approved credit.

Corley specializes in printing 8 1/2 x 11 books and catalogs (4-color on covers only). Their quality is good, and their service is wonderful. They can give you a fast turnaround time if needed. They provided us with one of the lowest quotes for 5000 copies of a 48-page 8 1/2 x 11 book; they even sent us a mock-up of the book -- a very nice touch!

Country Press
P O Box 489
Middleborough MA 02346

617-947-4485
Alfred E. Wolf
President

They specialize in producing bound galley proofs with a fast turnaround. They have not answered our last two RFQ's or printer surveys, so we cannot give you any further details.

Courier Graphics
4325 Old Shepherdsville Road
P O Box 18640
Louisville KY 40218

502-458-5303
R. Bruce Besten
President

They have never answered any of our RFQ's or survey forms. They apparently do not do short runs.

Crane Duplicating Service Inc.   617-362-2700
1611 Main Street / P O Box 487   Kenyon Gregiore
Barnstable MA 02630   Customer Service Director

Quantities: Min: 20   Max: 3000   Opt: 20 - 1500

Book Sizes: 5 1/2 x 8 1/2; 6 x 9; 8 1/2 x 11

Bindings: [X] PB   [X] SS   [ ] HC   [X] C/SB

Capabilities: [ ] Magazines   [X] Galley Copies
                [X] Journals   [ ] Demand Printing
                [X] Cookbooks   [X] 4-color Juvenile Books
                [ ] Yearbooks   [X] Annual Reports/Brochures
                [ ] Catalogs   [ ] Other Commercial Printing

Services: [X] Typesetting   [X] Teletypesetting
            [X] Design and Pasteup   [ ] Editing
            [ ] 4-color Printing   [ ] Warehousing/Shipping

Terms: 2% 10, net 30.

They were the first company to specialize in readers' proofs
(galley copies). They also do short-run book and journal
reprints. Their average time from receipt of copy to finished
books is "7 working days." They provide overnight shipping service to New York and Boston.

==

Crest Litho   518-456-2296 / 212-903-4740
2053 Central Ave / P O 12125   Christine Warger
Albany NY 12212   Customer Service Representative

Quantities: Min: 1000   Max: 500,000   Opt: 50,000 - 100,000

Book Sizes: Many sizes from 4 x 8 on up.

Bindings: [I] PB   [I] SS   [ ] HC   [I] C/SB

Capabilities: [X] Magazines   [ ] Galley Copies
                [X] Journals   [ ] Demand Printing
                [ ] Cookbooks   [ ] 4-color Juvenile Books
                [ ] Yearbooks   [X] Annual Reports/Brochures
                [X] Catalogs   [ ] Other Commercial Printing

Services: [ ] Typesetting   [ ] Teletypesetting
            [ ] Design and Pasteup   [ ] Editing
            [X] 4-color Printing   [X] Warehousing/Shipping

Crest Litho continued

Terms: Net 30 days.

   Crest is a multi-purpose book/commercial printer. "We service several different markets with the overall objective of being able to do most any kind of printing specification."

   User Comment: "They are my best printer! I've worked with them for over five years."

RATINGS	1	2	3	4	5	6	7	8	9	10	Ave	
Speed	-	-	-	-	-	-	1	-	-	-	---	1
Price	-	-	-	-	-	-	1	-	-	-	---	
Dependability	-	-	-	-	-	-	-	-	-	1	---	
Service . . .	-	-	-	-	-	-	-	-	-	1	---	
Quality . . .	-	-	-	-	-	-	-	-	-	1	---	
Overall . . .	-	-	-	-	-	-	-	-	-	1	---	

Cushing-Malloy
1350 N Main St / P O Box 8632
Ann Arbor MI 48107

313-663-8554
Thomas F. Weber
Vice President Sales

Quantities: Min: 100   Max: 15,000   Opt: 3000

Book Sizes: All sizes between 5 1/2 x 8 1/2 and 8 1/2 x 11.

Bindings: [I] PB   [I] SS   [O] HC   [O] C/SB

Capabilities: [ ] Magazines   [ ] Galley Copies
                [X] Journals   [ ] Demand Printing
                [ ] Cookbooks   [ ] 4-color Juvenile Books
                [ ] Yearbooks   [ ] Annual Reports/Brochures
                [ ] Catalogs   [ ] Other Commercial Printing

Services: [ ] Typesetting   [ ] Teletypesetting
            [ ] Design and Pasteup   [ ] Editing
            [ ] 4-color Printing   [X] Warehousing/Shipping

Terms: Flexible.

   Printers for over 35 years, they specialize in short to medium runs of books and journals. "We are small enough to adapt to our customers' requirements but large enough to meet the manufacturing needs of most small to medium-run publishers."

   See Cushing-Malloy's user rating following these ads.

RATINGS	1	2	3	4	5	6	7	8	9	10	Ave	
Speed	-	-	-	-	1	-	-	-	-	-	---	1
Price	-	-	1	-	-	-	-	-	-	-	---	
Dependability	-	-	-	-	1	-	-	-	-	-	---	
Service . . .	-	-	-	-	1	-	-	-	-	-	---	
Quality . . .	-	-	-	-	1	-	-	-	-	-	---	
Overall . . .	-	-	-	-	1	-	-	-	-	-	---	

Data Copi
2900 Peachtree Road NE
Atlanta   GA 30305

404-261-0133
Vicki Hodges
Vice President

Quantities: Min: 10   Max: 50,000   Opt: not stated

Book Sizes: 5 1/2 x 8 1/2; 6 x 9; 8 1/2 x 11

Bindings: [X] PB   [X] SS   [ ] HC   [X] C/SB

Capabilities: [ ] Magazines   [ ] Galley Copies
                [ ] Journals   [X] Demand Printing
                [ ] Cookbooks   [ ] 4-color Juvenile Books
                [ ] Yearbooks   [ ] Annual Reports/Brochures
                [ ] Catalogs   [ ] Other Commercial Printing

Services: [ ] Typesetting   [ ] Teletypesetting
            [ ] Design and Pasteup   [ ] Editing
            [ ] 4-color Printing   [ ] Warehousing/Shipping

Terms: Not stated.

Their speciality is electronic composition and publishing, but apparently do not do the typesetting themselves.

User Comment: "Initial quotes did not meet final price (much higher final price). Poor coordination (they were going through a management change at the time)."

RATINGS	1	2	3	4	5	6	7	8	9	10	Ave	
Speed	-	-	1	-	-	-	-	-	-	-	---	1
Price	-	-	-	-	1	-	-	-	-	-	---	
Dependability	-	-	-	1	-	-	-	-	-	-	---	
Service . . .	-	-	1	-	-	-	-	-	-	-	---	
Quality . . .	-	-	-	-	-	1	-	-	-	-	---	
Overall . . .	-	-	-	1	-	-	-	-	-	-	---	

Davidson Printing Company
120 N 2nd Avenue
Duluth MN 55802

218-727-8721
Attn: Sales Representative

They have never answered any of our RFQ's or survey forms.
They apparently do not do short runs.

Davis Printing Corporation
640 Dell Road
Carlstadt NJ 07072

201-935-5100 / 212-736-1980
Gilbert Davis
President

They also did not answer our recent printer survey form. They
apparently do not do short runs.

The Delmar Company
9601 Monroe Rd / PO Box 220025
Charlotte NC 28222

704-847-9801
Jim Hunter
General Manager

They were recommended to us by one of their customers, but
they did not answer our RFQ or printer survey. They offer complete typesetting and 4-color services as well as almost any
binding you could want.

User Comment: "Best on hardcover books and odd sizes."

RATINGS	1	2	3	4	5	6	7	8	9	10	Ave	
Speed	-	-	-	1	-	-	-	-	-	-	---	1
Price	-	-	-	-	-	-	-	1	-	-	---	
Dependability	-	-	-	-	-	-	-	1	-	-	---	
Service . . .	-	-	-	-	-	-	-	1	-	-	---	
Quality . . .	-	-	-	-	-	-	1	-	-	-	---	
Overall . . .	-	-	-	-	-	-	1	-	-	-	---	

```
* * * * * * * * * * * * * * * * * * * * * * * * * * * * * * * *
*  Tip:  You can obtain a two-color effect with only one color  *
*        by using screens, dropouts, and reverses.  We used     *
*        all three techniques to create the one-color cover on  *
*        the Second Edition of this Directory (and saved two    *
*        hundred dollars over the cost of a standard two-color  *
*        cover).  Plus, we feel the cover actually looked       *
*        better with one color than it would have with two.     *
* * * * * * * * * * * * * * * * * * * * * * * * * * * * * * * *
```

Delta Lithograph
14731 Califa Street
Van Nuys CA 91411-3119

213-873-4910
Ken Hoffman
Sales Manager

In California, call their toll-free number: 800-223-1478; in San Jose, call 818-781-2460.

Quantities: Min: 500   Max: open   Opt: 2000+

Book Sizes: Almost any size.

Bindings: [I] PB   [I] SS   [ ] HC   [ ] C/SB

Capabilities: [ ] Magazines   [ ] Galley Copies
                [X] Journals   [ ] Demand Printing
                [ ] Cookbooks   [ ] 4-color Juvenile Books
                [ ] Yearbooks   [X] Annual Reports/Brochures
                [ ] Catalogs   [ ] Other Commercial Printing

Services: [X] Typesetting   [X] Teletypesetting
            [X] Design and Pasteup [ ] Editing
            [ ] 4-color Printing   [X] Warehousing/Shipping

Terms: Net 30 days.

Delta has just been bought by the Bertelsmann Group (the German publishing group which also owns Offset Paperback Manufacturers and Bantam Books). Delta has produced over 25,000 different books, manuals, directories, and catalogs over the past 30 years. They have a reputation for superb quality and excellent service. They were among the top rated book printers in the 1981 Small Press Review survey, and were very highly rated by four users who responded to our recent ratings survey.

Delta's main drawback is that they tend to be more expensive than the Ann Arbor printers. Their prices are approximately 10 - 20% higher, but when shipping costs for West Coast publishers are taken into consideration, Delta does become competitive.

Send for their free Planning Guide, which presents excellent guidelines for preparing your camera-ready copy for the printer.

User Comments: "Good work, good price, deliver when they promise." ... "I trust them to do a good job, and they've never violated that trust or wasted my time, energy or money." ... "Mixed feelings about Delta, though they are generally reliable, but sometimes very slow." ... "These people were a pleasure to deal with, though their production manager was often difficult to reach. The rest were excellent. They get first crack at my next book."

See the ratings survey on the page following Delta's ad.

60

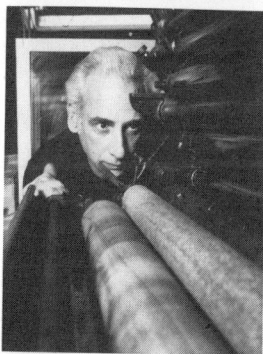

Delta Lithograph continued

RATINGS		1	2	3	4	5	6	7	8	9	10	Ave	
Speed		–	–	–	–	1	2	–	–	–	1	6.75	4
Price		–	–	–	1	–	2	–	–	1	–	6.25	
Dependability		–	–	–	1	–	–	–	–	–	3	8.5	
Service . . .		–	–	–	–	1	–	–	–	1	2	8.5	
Quality . . .		–	–	–	–	1	–	–	–	1	2	8.5	
Overall . . .		–	–	–	–	1	–	–	–	2	1	8.25	

Desaulniers Printing Company   309-799-7331
4905 - 77th Avenue   David E. Burt
Milan IL 61264   Sales Manager

   Their work seems to be of good quality, and they have done
short-run book printing in the past. However, they did not
answer our printers survey or RFQ, so they may no longer be doing
short-runs.

John Deyell Company   705-324-6148
107 Lindsay Street S   J. D. Beaudin
Lindsay, Ontario   Marketing Manager
K9V 2M5   Canada

   They are members of the Book Manufacturers Institute and have
been in business for many years, but they apparently do not do
short-runs since they did not answer our printer survey or RFQ.

Dickinson Press   616-451-2957
630 Myrtle Street NW   Vern De Weerd
Grand Rapids MI 49504   President

   They have a Chicago sales office: 312-668-1944.

Quantities: Min: 3000   Max: none   Opt: 10,000

Book Sizes: 5 1/2 x 8 1/2; 6 x 9

Bindings: [I] PB   [I] SS   [X] HC   [X] C/SB

Capabilities: [ ] Magazines   [ ] Galley Copies
                [ ] Journals   [ ] Demand Printing
                [X] Cookbooks   [X] 4-color Juvenile Books
                [ ] Yearbooks   [X] Annual Reports/Brochures
                [X] Catalogs   [X] Other Commercial Printing

Services: [ ] Typesetting   [ ] Teletypesetting
            [ ] Design and Pasteup [ ] Editing
            [X] 4-color Printing   [X] Warehousing/Shipping

Terms: 2% 10 days, net 30.

We do not know much more about this printer, but one of our
correspondents who has used them says they are "very good."

==

Dinner & Klein   206-682-2494
600 S Spokane St / PO Box 3814   Jenny L. Saxton
Seattle WA 98124   National Sales Coordinator

Quantities: Min: 1000   Max: 150,000   Opt: 50,000

Book Sizes: 5 1/4 x 8 1/2 and 8 1/2 x 11

Bindings: [I] PB   [I] SS   [ ] HC   [ ] C/SB   [I] glue binding

Capabilities: [ ] Magazines   [ ] Galley Copies
                [ ] Journals   [ ] Demand Printing
                [ ] Cookbooks   [ ] 4-color Juvenile Books
                [ ] Yearbooks   [X] Annual Reports/Brochures
                [X] Catalogs   [X] Other Commercial Printing

Services: [ ] Typesetting   [ ] Teletypesetting
            [ ] Design and Pasteup [ ] Editing
            [ ] 4-color Printing   [ ] Warehousing/Shipping

Terms: Payment in full with order.

Since 1948, Dinner & Klein has specialized in printing one or
two-color direct mail components and catalogs of various paper
grades (newsprint, 34 lb. Antique, etc.). They offer good prices
on self-cover catalogs and booklets using their standard paper
stocks.

Send for their standard price list which also includes a
wealth of excellent advice about preparing camera-ready copy and
photographs for printing. Note that they advertise nationally,
and are quite accustomed to dealing with customers by mail from
all parts of the United States and Canada.

R. R. Donnelley & Sons Company   800-428-0832   /   317-362-1300
Rt 32 West   Chuck Harpel
Crawfordsville IN 47933   Direct Sales Coordinator

R. R. Donnelley's corporate offices are located at 2223 King Drive, Chicago, IL 60606; phone: 312-362-8000.

They also have another short-run book production facility at 1400 Kratzer Road, Harrisonburg, VA 22801

Quantities: Min: 2000   Max: 100,000   Opt: 5000 - 10,000

Book Sizes: Any size between 5 1/2 x 8 1/2 to 7 3/8 x 9 1/4

Bindings: [I] PB   [I] SS   [I] HC   [I] C/SB

Capabilities: [X] Magazines   [ ] Galley Copies
                [X] Journals   [ ] Demand Printing
                [X] Cookbooks   [X] 4-color Juvenile Books
                [ ] Yearbooks   [ ] Annual Reports/Brochures
                [X] Catalogs   [ ] Other Commercial Printing

Services: [X] Typesetting   [ ] Teletypesetting
            [ ] Design and Pasteup   [ ] Editing
            [X] 4-color Printing   [X] Warehousing/Shipping

Other services: software packaging and computer disc replication

Terms: Net 30 with approved credit.

R. R. Donnelley is itself the largest printer in the United States, with an annual sales volume of $1,400,000,000. They print more telephone directories and Bibles than anyone else; they also print many national magazines.

The Crawfordsville and Harrisonburg plants are their short-run book printing facilities. Their sales coordinator says that they are very tenacious when it comes to meeting their schedule commitments; he also says that books are run in the order the jobs are received (that means that a small job won't get bumped by a larger job from a regular customer).

They can provide typesetting via an associated company but do not do typesetting at either printing plant.

Send for their Donnelley Guide to Book Planning (free with letterhead request). It includes many samples of paper and cover stock.

User Comment: "Prices cannot be beat but quality, compared to our other suppliers, is inferior." But, note on the next page, that another user rates their quality very highly.

RATINGS	1	2	3	4	5	6	7	8	9	10	Ave	
Speed	-	-	-	-	-	1	-	-	1	-	---	2
Price	-	-	-	-	-	-	1	-	-	1	---	
Dependability	-	-	-	-	-	-	-	1	1	-	---	
Service . . .	-	-	-	-	-	-	-	1	1	-	---	
Quality . . .	-	-	1	-	-	-	-	-	1	-	---	
Overall . . .	-	-	-	-	1	-	-	-	1	-	---	

Dynamic Printing
4221 S Avenida Paisano
Tucson AZ 85746

602-883-5610
J. A. Saxon
Owner

SP

Quantities: Min: 10   Max: open   Opt: 2000

Book Sizes: Any size book that will fit on an 11 x 17 press size

Bindings: [I] PB   [I] SS   [ ] HC   [I] C/SB

Capabilities: [X] Magazines   [ ] Galley Copies
                [X] Journals   [ ] Demand Printing
                [ ] Cookbooks   [ ] 4-color Juvenile Books
                [ ] Yearbooks   [X] Annual Reports/Brochures
                [ ] Catalogs   [X] Other Commercial Printing

Services: [ ] Typesetting   [ ] Teletypesetting
            [X] Design and Pasteup   [X] Editing
            [ ] 4-color Printing   [ ] Warehousing/Shipping

Terms: Not stated.

Dynamic, a family-owned business, publishes health books as Better Health Publications. They will also help others self-publish their own books. Their motto is, "Quality Comes First."

```
* * * * * * * * * * * * * * * * * * * * * * * * * * * * * * * * * *
*  Tip:  If you provide complete camera-ready copy to your       *
*        printer, you may not need to see pressproofs. Most      *
*        quality printers will reproduce your camera-ready       *
*        copy exactly as you provided it. Hence, you can save    *
*        the proof charges (which cost anywhere from $50 to      *
*        $250) and save the time that would otherwise be taken   *
*        up in sending the proofs back and forth for approval.   *
*        However, when working with a printer for the first      *
*        time, it is always safer to require press proofs.       *
*        Always check your press proofs very thoroughly.         *
* * * * * * * * * * * * * * * * * * * * * * * * * * * * * * * * * *
```

Eastern Lithographing
2815 N 17th Street
Philadelphia PA 19132

215-225-1150
David M. Silver
Vice President Sales

Quantities: Min: 200   Max: 80,000   Opt: 500 - 5000

Book Sizes: Any size from 3 1/2 x 5 1/2 to 18 1/2 x 14

Bindings: [X] PB   [X] SS   [X] HC   [X] C/SB

Capabilities: [ ] Magazines   [ ] Galley Copies
                [X] Journals   [ ] Demand Printing
                [ ] Cookbooks   [X] 4-color Juvenile Books
                [X] Yearbooks   [ ] Annual Reports/Brochures
                [X] Catalogs   [ ] Other Commercial Printing

Services: [ ] Typesetting   [ ] Teletypesetting
            [ ] Design and Pasteup   [ ] Editing
            [X] 4-color Printing   [ ] Warehousing/Shipping

Terms: 1% 20 days, net 30.

   Eastern offers more competitive prices on books with a high
page count because their press requires half as many plates and
half the makeready time of many other presses. They have two
other specialties: 4-color children's books and loose-leaf
manuals (which they can produce in 10 working days).

   They are a small company which emphasizes their service and
speed: "We view service as our real job, as well as printing."

═══

Eastern Press
654 Orchard St / P O Box 1650
New Haven CT 06507

203-777-2353
Ray A. Johnson
President

   Eastern specializes in printing 4-color art books, but we do
not know if they do short-runs. They failed to answer our last
two printer surveys.

═══

Eastern Publishing Graphics
24 Tannery Lane N
Weston CT 06680

?
Attn: Sales Representative

   These folks may well be out of business.

Eastwood Printing & Publishing
2901 Blake
Denver CO 80205

303-296-1905
Attn: Sales Representative

They are listed as book printers in the Denver phone book, but they have never answered our RFQ's or printer surveys. They may not be interested in doing short runs.

Economy Bookcraft
681 Market St #531
San Francisco CA 94105

415-362-2708
Evelyn Pavlik
Sales Representative

Economy is a typesetter and printing broker who can arrange to have books printed in any quantity from 250 to 100,000. They also offer distribution services in Northern California and direct mail program development.

Economy Printing Company
Route 50
Easton MD 21601

?
Luther Smith

Economy was used by one of the respondents to our book printer ratings survey (see comments and ratings below). However, they did not respond to our query for more information, so we cannot give you any further details about their capabilities.

User Comment: "Economy printed 5000 small booklets for us. Did an excellent job."

RATINGS	1	2	3	4	5	6	7	8	9	10	Ave	
Speed	-	-	-	-	-	-	1	-	-	-	---	1
Price	-	-	-	-	1	-	-	-	-	-	---	
Dependability	-	-	-	-	-	-	-	-	-	1	---	
Service . . .	-	-	-	-	-	-	-	1	-	-	---	
Quality . . .	-	-	-	-	-	-	-	-	-	1	---	
Overall . . .	-	-	-	-	-	-	-	1	-	-	---	

```
* * * * * * * * * * * * * * * * * * * * * * * * * * * * * *
*  Tip:  When doing two or three-color printing, avoid close   *
*        registrations.  They require extra prep time and can  *
*        result in a higher reject rate as well.               *
* * * * * * * * * * * * * * * * * * * * * * * * * * * * * *
```

Edison Lithographing
418 W 25th Street
New York NY 10001

212-741-2212
I. Gross
President

They have never answered any of our RFQ's or survey forms. They apparently do not do short runs (although they are listed as doing short runs in LMP).

Edwards & Broughton
1821 North Boulevard
Raleigh NC 27611

919-833-6601
Alan Phillips
Sales Manager

Quantities: Min: 2500   Max: 250,000   Opt: 25,000 - 100,000

Book Sizes: any size from 5 1/2 x 8 1/2 to 8 1/2 x 11

Bindings: [I] PB   [I] SS   [O] HC   [I] C/SB

Capabilities: [X] Magazines   [ ] Galley Copies
                [ ] Journals   [ ] Demand Printing
                [X] Cookbooks   [ ] 4-color Juvenile Books
                [ ] Yearbooks   [X] Annual Reports/Brochures
                [X] Catalogs   [X] Other Commercial Printing

Services: [X] Typesetting   [X] Teletypesetting
            [X] Design and Pasteup   [ ] Editing
            [X] 4-color Printing   [ ] Warehousing/Shipping

Terms: Net 30 with approved credit. Special arrangements may be considered as long as your credit is good.

They can also print brochures, folders, large 4-color posters (as large as 43" x 60"), and point of purchase displays. Annual reports can only be done on their sheet-fed presses, which have a maximum capacity of 25,000 copies.

```
* * * * * * * * * * * * * * * * * * * * * * * * * * * * * * * *
*  Tip:  With some printers you will be able to save money by   *
*        providing your camera-ready copy on single sheets      *
*        that are the same size as the finished book page.      *
*        Others require you to use their special layout pages.  *
*        Check first to find out what they want from you.       *
*        In either case, the printer has a system that allows   *
*        a number of pages to be shot by the camera at the      *
*        same time, thus cutting prep costs.  It usually also   *
*        makes the job of stripping negatives easier for them.  *
* * * * * * * * * * * * * * * * * * * * * * * * * * * * * * * *
```

Edwards Brothers
2500 S State St / P O Box 1007
Ann Arbor MI 48106-1007

313-769-1000
Donald G. Ford
National Sales Manager

They have sales offices all over the place:

617-863-8787	Boston MA	212-867-2830	New York NY
312-787-9206	Chicago IL	609-235-4164	Philadelphia PA
216-831-9562	Cleveland OH	415-828-2377	San Francisco CA
609-235-4164	Mount Laurel NH	301-587-7280	Washington DC
01-852-7967	London, England		

They also have another printing facility at 800 Edwards Drive, P. O. Box 1025, Lillington, NC 27546; phone: 919-893-2717.

Quantities: Min: 500   Max: 20,000   Opt: 4,000

Book Sizes: Almost any size.

Bindings: [I] PB   [I] SS   [I] HC   [I] C/SB

Capabilities: [ ] Magazines   [ ] Galley Copies
                 [X] Journals   [ ] Demand Printing
                 [ ] Cookbooks   [ ] 4-color Juvenile Books
                 [ ] Yearbooks   [ ] Annual Reports/Brochures
                 [ ] Catalogs   [ ] Other Commercial Printing

Services: [X] Typesetting   [ ] Teletypesetting
            [ ] Design and Pasteup   [ ] Editing
            [ ] 4-color Printing   [X] Warehousing/Shipping

Terms: Net 30.

Edwards Brothers, the grand-daddy of the Ann Arbor short-run book printers (since 1893), is one of the 100 largest printers in the U.S. (with over $25,000,000 in business per year). They were rated among the top ten short-run printers in the U.S. in the 1981 Small Press Review survey and were also highly rated by many of those who answered our ratings survey this year, though several people have had problems working with them (see the comments below and on the next page). They tend to be very slow.

They can print 4-color covers but only one or two-color text.

User Comments: "Good quality and versatile. Most often comes in second in quotes, very close to best quote." ... "Best paper selection for the small publisher. Sales rep is more helpful than their in-house estimators/service people." ... "A little slower than most, but reliable." ... "Bad experience. They designed the book too small for comb binding, and they printed one page with a line missing, even though the line showed in the proof."   ... more comments on the next page ...

Edwards Brothers continued

User Comments: "Our favorite; their quality is hardly 'fine' and their speed is slow but, they are best in every other category." ... "Always good for quality, and recently their prices have been very competitive." ... "The only reason I still do business with these guys is that I refuse to believe anything else can go wrong. Their prices are good, but I've had binding problems, imposition problems, billing problems, printing problems, coating problems, you name it. I've heard good things about them from others, and their paper stocking program is as good as Fairfield's, maybe even better, but I've had too many bad experiences to feel comfortable with them."

RATINGS	1	2	3	4	5	6	7	8	9	10	Ave	
Speed	-	1	2	2	2	1	-	-	-	-	4.0	8
Price	-	-	-	-	-	-	1	5	1	1	8.25	
Dependability	1	-	1	-	-	1	2	-	1	2	6.63	
Service . . .	-	-	-	-	1	1	1	-	3	2	8.13	
Quality . . .	-	1	1	-	-	-	2	1	1	2	7.0	
Overall . . .	-	-	1	1	-	-	2	2	1	1	7.0	

Eerdmans Printing Company   616-451-0763
231 Jefferson Avenue SE   Don Miller
Grand Rapids MI 49503   Assistant Plant Manager

Quantities: Min: 500   Max: 400,000   Opt: 20,000

Book Sizes: any size from 4 x 6 to 8 1/2 x 11

Bindings: [I] PB   [I] SS   [I] HC   [ ] C/SB

Capabilities: [ ] Magazines   [ ] Galley Copies
                [ ] Journals   [ ] Demand Printing
                [X] Cookbooks   [ ] 4-color Juvenile Books
                [ ] Yearbooks   [ ] Annual Reports/Brochures
                [X] Catalogs   [ ] Other Commercial Printing

Services: [ ] Typesetting   [ ] Teletypesetting
            [ ] Design and Pasteup [ ] Editing
            [X] 4-color Printing   [X] Warehousing/Shipping

Terms: 1% 10, net 30 with approved credit.

They have been printing books for over 50 years as the printing arm of Eerdmans Publishing Company. For more information, see the page following their ad.

They print over 1000 titles (including low or high quantities of mass market paperbacks) every year. Their customers include major publishers as well as independents. They emphasize their customer service and will work with you to produce the best book possible. We found them very accommodating in printing the Second Edition of this Directory, though one user of the Second Edition did note that Eerdmans bound the book cross-grain (which causes the pages in the book to be less flexible and, hence, not stand open as easily).

User Comment: "A very satisfactory company with common, sensible and interesting managers."

RATINGS	1	2	3	4	5	6	7	8	9	10	Ave	
Speed	-	-	-	-	-	-	-	1	-	1	---	2
Price	-	-	-	-	-	-	-	-	1	1	---	
Dependability	-	-	-	-	-	-	-	1	-	1	---	
Service . . .	-	-	-	-	-	-	-	-	1	1	---	
Quality . . .	-	-	-	-	-	-	1	-	-	1	---	
Overall . . .	-	-	-	-	-	-	-	1	-	1	---	

Evangel Press
301 N Elm Street
Nappanee IN 46550

219-773-3164
Jon Stepp
Manager

Quantities: Min: 500   Max: 20,000   Opt: 5000 - 10,000

Book Sizes: 5 1/2 x 8 1/2; 6 x 9; 8 1/2 x 11

Bindings: [I] PB   [I] SS   [O] HC   [O] C/SB

Capabilities: [ ] Magazines   [ ] Galley Copies
               [X] Journals   [ ] Demand Printing
               [X] Cookbooks   [ ] 4-color Juvenile Books
               [ ] Yearbooks   [ ] Annual Reports/Brochures
               [ ] Catalogs   [X] Other Commercial Printing

Services: [X] Typesetting   [ ] Teletypesetting
           [X] Design and Pasteup   [X] Editing
           [X] 4-color Printing   [ ] Warehousing/Shipping

Terms: Net 30 days with established credit.

They have been in business for over 60 years printing books and other items primarily for local businesses and publishers.

Evergreen Press
1070 SE Marine Drive
Vancouver, British Columbia
V5X 2V4   Canada

604-321-2231
Warren Marrs
Sales Representative

They did not answer our RFQ or printer survey, so we cannot give any details regarding their capabilities or services. They probably do not do short runs.

Exposition Press of Florida
1701 Blount Road #C
Pompano Beach FL 33069

305-979-3200
Edward Uhlan
President

SP

Quantities: Min: 100   Max: 100,000   Opt: 1000 - 5000

Book Sizes: any and all sizes

Bindings: [X] PB   [ ] SS   [X] HC   [X] C/SB

Capabilities: [ ] Magazines   [ ] Galley Copies
                 [ ] Journals   [ ] Demand Printing
                 [X] Cookbooks   [X] 4-color Juvenile Books
                 [ ] Yearbooks   [ ] Annual Reports/Brochures
                 [ ] Catalogs   [ ] Other Commercial Printing

Services: [X] Typesetting   [X] Teletypesetting
             [X] Design and Pasteup   [X] Editing
             [X] 4-color Printing   [X] Warehousing/Shipping

Other Services: Marketing and promotional help.

Terms: Not stated, but probably prepayment is required.

Edward Uhlan has been involved in vanity book publishing for over 49 years. As a vanity publisher, Exposition also provides limited marketing and promotional help.

Fabe Litho Ltd.
602 W Rillito Street
Tucson AZ 85705

602-622-2857
Hollis Z. Fabe
Vice President

They asked to be listed in this Directory but were too late to get a full listing. You might want to query them to see if they can meet your needs (especially if you're located nearby).

Faculty Press
1449 - 37th Street
Brooklyn NY 11218

718-851-6666
Walter Heitner
Vice President Sales

Quantities: Min: 500   Max: 35,000   Opt: 3000 - 5000

Book Sizes: They can print any size, including many odd sizes.

Bindings: [I] PB   [I] SS   [I] HC   [I] C/SB

Capabilities: [X] Magazines   [ ] Galley Copies
               [X] Journals   [ ] Demand Printing
               [X] Cookbooks   [X] 4-color Juvenile Books
               [ ] Yearbooks   [X] Annual Reports/Brochures
               [X] Catalogs   [X] Other Commercial Printing

Services: [X] Typesetting   [ ] Teletypesetting
           [X] Design and Pasteup   [ ] Editing
           [X] 4-color Printing   [ ] Warehousing/Shipping

Terms: Net 30 with approved credit.

   Don't let their name fool you -- they are not scholarly book
printers. They print posters, catalogs, and inserts as well as
high quality trade books. They are most price competitive when
printing odd-sized books and books requiring high quality (photos
and/or color). They can provide rapid turnaround when required
(for example, a 300 page perfectbound book in under two weeks).
"We're book fans" and "friends of independent publishers"

===

Fairfield Graphics
P O Drawer AN
Fairfield PA 17320

717-642-5871
Eddie Owens
Marketing Manager

   Fairfield is a division of Arcata, which has a toll-free phone
number you can call for quotes: 800-722-7020.

Quantities: Min: 1500   Max: open   Opt: 10,000 - 25,000

Book Sizes: 5 1/2 x 8 1/2; 6 x 9

Bindings: [I] PB   [ ] SS   [I] HC   [ ] C/SB

Capabilities: [ ] Magazines   [ ] Galley Copies
               [ ] Journals   [ ] Demand Printing
               [ ] Cookbooks   [ ] 4-color Juvenile Books
               [ ] Yearbooks   [ ] Annual Reports/Brochures
               [ ] Catalogs   [ ] Other Commercial Printing

Services: [ ] Typesetting   [ ] Teletypesetting
           [ ] Design and Pasteup [ ] Editing
           [ ] 4-color Printing   [ ] Warehousing/Shipping

Terms: Negotiable.

   Their speciality is 1 or 2-color trade books and college textbooks. Even though they are a part of the Arcata book group,
they will often compete for business against the other printers
in the group (Halliday Litho and Kingsport Press).

   User Comment: "Fairfield is the newest of Arcata's plants,
and functions as an independent subsidiary of Arcata. It is a
state of the art printing facility. They only have web printing
equipment, but they still can help the short-run buyer. ... They
can be very fast, and very good. The best thing about them is
their stocking program of papers. They have a very wide
selection of paper. ... The worst part of FG, though, is that
they are geared to dealing with huge publishers, and will follow
your instructions to the letter. Before you work with them, you
have to know how their entire process works -- the vocabulary,
the materials, the press equipment, everything -- or you'll end
up with a real problem. ... When they're good, they're very,
very good. When they're bad, you're in trouble."

RATINGS	1	2	3	4	5	6	7	8	9	10	Ave	
Speed	–	–	–	1	–	1	–	–	–	–	---	2
Price	–	–	–	–	1	–	1	–	–	–	---	
Dependability	–	–	–	–	–	✓	1	1	–	–	---	
Service . . .	–	–	1	–	–	–	–	–	1	–	---	
Quality . . .	–	–	–	1	–	–	–	1	–	–	---	
Overall . . .	–	–	–	–	1	–	–	1	–	–	---	

Fairview Litho   914-473-4747
72 Fairview Avenue   Patrick Pacio
Poughkeepsie NY 12601   Vice President Sales & Marketing

   They have never answered any of our RFQ's or survey forms.
Although they are listed as short-run book printers in the LMP,
they do not seem to be very responsive.


```
* * * * * * * * * * * * * * * * * * * * * * * * * * * * * * *
*  Tip:  Avoid special requests (odd sizes, unusual papers,    *
*        special effects) unless they contribute to the        *
*        content of the book.  Extras cost time and money.     *
* * * * * * * * * * * * * * * * * * * * * * * * * * * * * * *
```

Fay Printing Center　　　　　　　417-883-1520
1923 S National / P O Box 3373　Eric Wittle
Springfield MO 65808　　　　　　Sales Representative

Quantities:　Min: 10　　Max: 10,000　　Opt: 1500 - 2500

Book Sizes:　the regular sizes plus several others

Bindings:　[O] PB　　[I] SS　　[O] HC　　[I] C/SB

Capabilities:　[X] Magazines　　　[ ] Galley Copies
　　　　　　　　[X] Journals　　　　[X] Demand Printing
　　　　　　　　[X] Cookbooks　　　[ ] 4-color Juvenile Books
　　　　　　　　[ ] Yearbooks　　　[X] Annual Reports/Brochures
　　　　　　　　[X] Catalogs　　　　[X] Other Commercial Printing

Services:　[X] Typesetting　　　　　[X] Teletypesetting
　　　　　　[X] Design and Pasteup　[ ] Editing
　　　　　　[X] 4-color Printing　　　[X] Warehousing/Shipping

Terms:　2% 10, net 30 days with approved credit.

　Fay, which has been in business for over 25 years, has two divisions: quick print and commercial. The commercial division can handle books, journals, catalogs, and so on. They'd like you to know that "service is really important to us."

William Feathers / Printers　216-774-1500
235 Artino Street　　　　　　　William Feathers Jr.
Oberlin OH 44074　　　　　　　President

　They can print multicolor pamphlets, but apparently are not interested in doing short-runs of books. They did not answer our most recent printer survey.

Federated Lithographers　　401-781-8100
369 Prairie Avenue　　　　　　Milton Walberg
Providence RI 02901　　　　　　Vice President Sales & Marketing

　They won the 1984 PIA (Printing Industries of America) Graphic Arts award for printing of magazines and house organs (three or more colors). They are members of the Book Manufacturers Institute. 10,000 seems to be their minimum print run. They have never answered any of our RFQ's or printer survey forms.

Fleetwood Graphics
588 Grand Canyon
Madison WI 53719

608-829-3536
Kenneth C. Flee Jr.
President

They provide typographic and design services and, according to their listing in LMP, also specialize in short-run pamphlets and perfectbound books; yet they have never answered any of our RFQ's or printer survey forms.

═══════════════════════════════════════

Foote & Davies - San Francisco
123 S Hill Drive
Brisband CA 94005

415-467-7100
Robert Burd
Sales Representative

Foote & Davies, a subsidiary of J. P. Stevens, is one of the 15 largest printers in the United States with annual sales of $150,000,000. Apparently they do not do short runs since they did not answer our recent printer survey.

═══════════════════════════════════════

Fort Orange Press
31 Sand Creek Rd / P O Box 828
Albany NY 12201

518-489-3233 / 800-448-4468 NY
Michele Dott
Estimator

Quantities: Min: 500   Max: 50,000   Opt: 5000 - 25000

Book Sizes: almost any size

Bindings: [I] PB   [I] SS   [I] HC   [I] C/SB

Capabilities: [X] Magazines   [X] Galley Copies
                [X] Journals   [ ] Demand Printing
                [X] Cookbooks   [X] 4-color Juvenile Books
                [ ] Yearbooks   [X] Annual Reports/Brochures
                [X] Catalogs   [X] Other Commercial Printing

Services: [X] Typesetting   [X] Teletypesetting
            [X] Design and Pasteup   [X] Editing
            [X] 4-color Printing   [X] Warehousing/Shipping

Terms: Net 30 days with credit approval.

The samples of their work that they sent us looked very good, although they were rated as being of "average quality" by one of the respondents to our ratings survey (see their ratings on next page).

Fort Orange Press continued

RATINGS	1	2	3	4	5	6	7	8	9	10	Ave
Speed	–	–	–	–	–	1	–	–	–	–	---	1
Price	–	–	–	–	–	1	–	–	–	–	---
Dependability	–	–	–	–	–	1	–	–	–	–	---
Service . . .	–	–	–	–	–	1	–	–	–	–	---
Quality . . .	–	–	–	–	1	–	–	–	–	–	---
Overall . . .	–	–	–	–	–	1	–	–	–	–	---

Four Corners Press   616-243-2015
2056 College SE   Charlotte Ellison
Grand Rapids MI 49507   President

   According to their ads, they specialize in design, editorial, composition, and production services for small to medium-sizes publishers and learned societies; however, they did not answer our RFQ or printer survey.

Four Winds Press   509-633-2060   SP
301 Lincoln   Attn: President
Coulee Dam WA 99116

   Four Winds was a small family-owned business serving self-publishing poets. However, they have not answered our last two printer surveys. They many no longer be in business.

Franklin Press   312-648-1512   SP
210 S Des Plaines   Attn: Sales Representative
Chicago IL 60606

   Franklin has also not answered our last two printer surveys. When I tried to call them, I got an answering service. They do books for self-publishers, small businesses, and organizations.

   User Comment: "Ten working days turnaround for magazines -- good on 4-color work." Is this the same Franklin Press, or another? We're not sure since the user comment was sent to us anonymously, and we were not able to follow it up.

RATINGS	1	2	3	4	5	6	7	8	9	10	Ave	
Speed	-	-	-	-	-	-	-	1	-	-	---	1
Price	-	-	-	-	-	1	-	-	-	-	---	
Dependability	-	-	-	-	-	1	-	-	-	-	---	
Service . . .	-	-	-	-	1	-	-	-	-	-	---	
Quality . . .	-	-	-	-	-	-	1	-	-	-	---	
Overall . . .	-	-	-	-	-	-	1	-	-	-	---	

Ray Freiman and Company
184 Brookdale Road
Stamford CT 06903

203-322-2474
Ray Freiman
President

They did not answer our RFQ's or printer surveys. They may be brokers for printing services rather than printers.

Friesen Printers
P O Box 720
Altona, Manitoba
R0G 0B0   Canada

204-324-6401
David Friesen Jr
President

They also have sales offices in the following areas:

403-253-3232	Calgary	306-652-6010	Saskatoon
403-452-6312	Edmonton	416-364-8747	Toronto
604-860 4442	Kelowna	604-576-2981	Vancouver
306-352-7954	Regina	204-475-9077	Winnipeg
902-895-8400	Maritime Provinces		

Their U.S. address is P. O. Drawer B, Neche, ND 58265.

Quantities: Min: 3000   Max: 30,000   Opt: 7000

Book Sizes: any book size fitting a 28" X 40" press sheet

Bindings: [X] PB   [X] SS   [X] HC   [X] C/SB

Capabilities:
[ ] Magazines   [ ] Galley Copies
[ ] Journals   [ ] Demand Printing
[X] Cookbooks   [X] 4-color Juvenile Books
[X] Yearbooks   [ ] Annual Reports/Brochures
[X] Catalogs   [ ] Other Commercial Printing

More details on the next page.

Friesen Printers continued

Services: [X] Typesetting   [ ] Teletypesetting
           [ ] Design and Pasteup   [ ] Editing
           [X] 4-color Printing   [X] Warehousing/Shipping

Terms: Net 30 days.

Only 5 miles from the U.S. border, they are quite capable of handling jobs from U.S. customers.

Fundcraft   800-351-7822 / 800-325-1994 TN
P O Box 340   Attn: Sales Representative
Collierville TN 38017

Fundcraft specializes in the production of personalized standard format cookbooks for fundraisers. Their terms are great: no downpayment, no finance charges. Their prices, though, are higher than what you could get from a good short-run printer. But, then, they have done all the work of designing the book for you and will do the typesetting as well. Send for their free fundraising kit.

Futura Printing   305-734-0825   SP
P O Drawer 99   Bob Steinmetz
Boynton Beach FL 33425-0099   President

Quantities: Min: 100   Max: 100,000   Opt: not stated

Book Sizes: 5 1/2 x 8 1/2; 6 x 9; 8 1/2 x 11

Bindings: [X] PB   [X] SS   [X] HC   [X] C/SB

Capabilities: [X] Magazines   [X] Galley Copies
                [X] Journals   [X] Demand Printing
                [X] Cookbooks   [X] 4-color Juvenile Books
                [ ] Yearbooks   [X] Annual Reports/Brochures
                [X] Catalogs   [X] Other Commercial Printing

Services: [X] Typesetting   [ ] Teletypesetting
            [X] Design and Pasteup [X] Editing
            [X] 4-color Printing   [ ] Warehousing/Shipping

Terms: 1/2 down, balance on delivery.

Futura prints books, magazines, price lists, newspapers and general commercial printing. Books for self-publishers are one of their specialties.

==

Ganis & Harris
260 Fifth Avenue
New York NY 10001

212-684-0850
Jim Harris
Vice President

Quantities: Min: 2500   Max: 100,000   Opt: 10,000

Book Sizes: many sizes from 4 1/4 x 6 to 11 x 14

Bindings: [I] PB   [I] SS   [I] HC   [I] C/SB

Capabilities: [ ] Magazines   [ ] Galley Copies
               [X] Journals   [ ] Demand Printing
               [ ] Cookbooks   [ ] 4-color Juvenile Books
               [ ] Yearbooks   [ ] Annual Reports/Brochures
               [ ] Catalogs   [ ] Other Commercial Printing

Services: [X] Typesetting   [ ] Teletypesetting
           [X] Design and Pasteup [X] Editing
           [X] 4-color Printing   [ ] Warehousing/Shipping

Terms: Net 30 days after establishing credit.

In business for over 35 years, they specialize in detailed, heavily illustrated books and technical books.

==

General Offset Company
234 16th Street
Jersey City NJ 07302

201-420-0500 / 212-925-1700
Murray A. Berger
President

Quantities: Min: 500   Max: 40,000   Opt: 15,000

Book Sizes: almost any size including all regular sizes

Bindings: [O] PB   [O] SS   [O] HC   [O] C/SB

Capabilities: [ ] Magazines   [ ] Galley Copies
               [ ] Journals   [ ] Demand Printing
               [ ] Cookbooks   [X] 4-color Juvenile Books
               [ ] Yearbooks   [ ] Annual Reports/Brochures
               [ ] Catalogs   [ ] Other Commercial Printing

Services: [ ] Typesetting   [ ] Teletypesetting
           [ ] Design and Pasteup   [ ] Editing
           [X] 4-color Printing   [ ] Warehousing/Shipping

Terms: Net 30 with approved credit.

   They do no bindings in-house; all are subcontracted out. They
specialize in color printing and halftone inserts. They do a lot
of 4-color children's books.

George Lithograph   415-397-2400
650 Second Street   George Calmenson
San Francisco CA 94107   Vice President of Marketing

Quantities: Min: 500   Max: 25,000   Opt: 10,000

Book Sizes: 5 1/2 x 8 1/2; 6 x 9

Bindings: [I] PB   [I] SS   [O] HC   [I] C/SB

Capabilities: [ ] Magazines   [ ] Galley Copies
               [X] Journals   [ ] Demand Printing
               [ ] Cookbooks   [ ] 4-color Juvenile Books
               [ ] Yearbooks   [ ] Annual Reports/Brochures
               [X] Catalogs   [X] Other Commercial Printing

Services: [X] Typesetting   [X] Teletypesetting
           [X] Design and Pasteup   [ ] Editing
           [ ] 4-color Printing   [X] Warehousing/Shipping

Terms: Net 30 with approved credit.

   They specialize in printing manuals and other computer docu-
mentation.

```
* * * * * * * * * * * * * * * * * * * * * * * * * * * * * * * *
*  Tip:  Design your books to fit the press (in signatures of  *
*        4, 8, 16, or 32 pages).  A book of 158 pages will     *
*        usually cost as much or more than a book of 160       *
*        pages because of additional labor charges in handling *
*        the incomplete signature.  Add several empty pages    *
*        or, better yet, use those extra pages to advertise    *
*        some of the other books you publish.  Or just add a   *
*        coupon so people can order additional copies of the   *
*        book itself (for themselves or for friends).          *
* * * * * * * * * * * * * * * * * * * * * * * * * * * * * * * *
```

Germac Printing
207 Tigard Plaza / P O 23877
Tigard OR 97223

503-639-0898
Cora Hall
Customer Service Manager

Quantities: Min: 10   Max: 40,000   Opt: 1000 - 10,000

Book Sizes: 5 1/2 x 8 1/2; 6 x 9; 8 1/2 x 11

Bindings: [O] PB   [I] SS   [I] HC   [O] C/SB

Capabilities: [ ] Magazines   [ ] Galley Copies
                [ ] Journals   [ ] Demand Printing
                [ ] Cookbooks   [ ] 4-color Juvenile Books
                [ ] Yearbooks   [X] Annual Reports/Brochures
                [X] Catalogs   [X] Other Commercial Printing

Services: [X] Typesetting   [ ] Teletypesetting
            [X] Design and Pasteup   [ ] Editing
            [ ] 4-color Printing   [ ] Warehousing/Shipping

Terms: Net 10 with approved credit.

A commercial printer with full graphics capability, Germac
specializes in one or two-color work with a maximum sheet size of
15" X 18".

Geryon Press   607-693-1572   SP
P O Box 770   Stuart McCarty
Tunnel NY 13848   Owner

Quantities: Min: 10   Max: 1000   Opt: 250 - 500

Book Sizes: 5 1/2 x 8 1/2; 6 x 9

Bindings: [I] PB   [I] SS   [O] HC   [O] C/SB

Capabilities: [ ] Magazines   [X] Galley Copies
                [ ] Journals   [X] Demand Printing
                [X] Cookbooks   [ ] 4-color Juvenile Books
                [ ] Yearbooks   [X] Annual Reports/Brochures
                [X] Catalogs   [X] Other Commercial Printing

Services: [X] Typesetting   [ ] Teletypesetting
            [X] Design and Pasteup   [ ] Editing
            [ ] 4-color Printing   [ ] Warehousing/Shipping

Terms: To be arranged individually.

... more info on next page

Geryon Press continued

Geryon is a small letterpress shop specializing in typesetting and printing poetry books, postcards, and broadsides. Their prices include typesetting, design, and printing (plus page and cover proofs).

Giant Horse and Company   415-468-0573
400 Talbert Street   Jeannie Ma
Daly City CA 94014   General Manager

Quantities: Min: 100   Max: 10,000   Opt: 500

Book Sizes: regular sizes between 4 1/2 x 5 1/2 and 9 x 12

Bindings: [I] PB   [I] SS   [O] HC   [I] C/SB

Capabilities: [ ] Magazines   [ ] Galley Copies
                 [X] Journals   [ ] Demand Printing
                 [ ] Cookbooks   [ ] 4-color Juvenile Books
                 [ ] Yearbooks   [X] Annual Reports/Brochures
                 [X] Catalogs   [X] Other Commercial Printing

Services: [X] Typesetting   [ ] Teletypesetting
             [ ] Design and Pasteup   [ ] Editing
             [X] 4-color Printing   [ ] Warehousing/Shipping

Terms: 50% down, 50% C.O.D.

Gilliland's Printing Company   800-332-8200 / 316-442-0500
215 N Summit St / P O Box 1107   Ed Gilliland
Arkansas City KS 67005   President

Quantities: Min: 1000   Max: 30,000   Opt: 1000 - 15,000

Book Sizes: 5 1/2 x 8 1/2; 6 x 9; 8 1/2 x 11

Bindings: [I] PB   [ ] SS   [ ] HC   [I] C/SB

Capabilities: [ ] Magazines   [ ] Galley Copies
                 [I] Journals   [ ] Demand Printing
                 [ ] Cookbooks   [ ] 4-color Juvenile Books
                 [ ] Yearbooks   [ ] Annual Reports/Brochures
                 [I] Catalogs   [ ] Other Commercial Printing

Services: [X] Typesetting   [X] Teletypesetting
           [X] Design and Pasteup   [ ] Editing
           [X] 4-color Printing   [X] Warehousing/Shipping

Terms: Net 30 days.

They specialize in printing and binding paperback books, directories, and college catalogs -- they are perhaps the largest printer of college catalogs in the country (over 300 per year).

===

Golden Bell Press   303-296-1600
2403 Champa   Attn: President
Denver CO 80205

They are listed in the Denver phone book as being capable of printing books, but they have never answered any of our RFQ's or survey forms. They apparently do not do short runs.

===

Golden Horn Press   415-845-4355
2506 Shattuck Avenue   Attn: Sales Representative
Berkeley CA 94704

They specialize in printing smaller books (4 x 7 or 5 1/2 x 8 1/2) in quantities from 300 to 10,000 copies. We have no other details regarding their services and capabilities.

===

Goshen Litho   914-469-2102
Route 17M   Harry Traub
Chester NY 10918   President

They print both paperbacks and magazines. We cannot give any other details regarding their services because they have never answered any of our printer surveys. They may not do short-runs.

```
* * * * * * * * * * * * * * * * * * * * * * * * * * * * * * * * *
*  Tip:  Build up your credit rating by paying your bills on    *
*        time.  Also, take advantage of any discounts for       *
*        paying early (you can save from 1% to as much as 5%).   *
* * * * * * * * * * * * * * * * * * * * * * * * * * * * * * * * *
```

Graphic Litho Corporation   617-683-2766
130 Shepard St / Industrial Pk   Ralph Wilbur
Lawrence   MA 01843   Sales Representative

They specialize in printing children's books and textbooks. We cannot give any further details regarding their capabilities because they have never answered any of our printer surveys.

Graphic Offset   301-539-8306
309 E Saratoga Street   President
Baltimore   MD 21202

They were recommended by a fellow publisher, but they did not respond to the printer surveys we sent them. They may not do short runs.

Great Northern Design   312-674-4740
5401 Fargo Avenue   Sidney M. Flom
Skokie   IL 60077   President

They have not responded to our last two RFQ's or printer surveys. The one quote we did receive from them was the highest for that particular book (1000 copies for $2695, as compared to the average quote of $1063 and the low quote of $540).

Griffin Printing & Lithograph   213-245-3671 / 818-244-2128
544 W Colorado Street   Jody Thompson
Glendale   CA 91204   Sales Manager

Quantities: Min: 500   Max: open   Opt: 5000

Book Sizes: 5 1/2 x 8 1/2, 6 x 9; 8 3/8 x 10 7/8

Bindings: [I] PB   [I] SS   [O] HC   [I] C/SB

Capabilities: [X] Magazines   [ ] Galley Copies
                [X] Journals   [ ] Demand Printing
                [X] Cookbooks   [ ] 4-color Juvenile Books
                [ ] Yearbooks   [X] Annual Reports/Brochures
                [X] Catalogs   [X] Other Commercial Printing

Services: [ ] Typesetting   [ ] Teletypesetting
           [ ] Design and Pasteup [ ] Editing
           [X] 4-color Printing   [X] Warehousing/Shipping

Terms: 2% 14, net 30 days with approved credit.

   They specialize in printing large size (8 3/8" x 10 7/8")
directories, manuals, and catalogs.

==

GRT Book Printing   415-534-5032
3960 E 14th Street   Ellen Geisler
Oakland CA 94601   Partner

Quantities: Min: 100   Max: 2000   Opt: 200 - 1000

Book Sizes: 5 1/2 x 8 1/2; 8 1/2 x 11

Bindings: [I] PB   [I] SS   [ ] HC   [I] C/SB

Capabilities: not stated

Services: not stated

Terms: 50% down, balance C.O.D.

   GRT specializes in print runs under 1000 copies. They serve
only customers from California. They publish a quarterly newsletter Printips that's supposed to have some good tips. Ask them
to send you a copy.

==

Guinn Printing Company   201-659-9000 / 212-924-0842
70 Hudson Street   Robert Guinn
Hoboken NJ 07030   President

   They did not answer our recent printer survey so we cannot
give you any details concerning their capabilities or services.

```
* * * * * * * * * * * * * * * * * * * * * * * * * * * * * * * *
*  Tip:  A lighter weight paper can save you money both on the  *
*        cost of the paper itself and in postage for mailing    *
*        the book.  Many 50 lb. or 55 lb. papers are now as     *
*        opaque and as durable as many 60 lb. papers.           *
* * * * * * * * * * * * * * * * * * * * * * * * * * * * * * * *
```

Haddon Craftsmen   717-348-9211
Ash St & Wyoming Avenue   Edward G. Rossi
Scranton PA 18509   National Sales Manager

Haddon has sales offices in Boston MA (617-426-1150), New York NY (212-533-9000), and San Francisco (415-391-8075). They have plants in four Pennsylvania locations: Allentown, Bloomsburg, Dunmore, and Scranton.

Haddon is one of the top 100 printers in the United States with annual sales of over $35,000,000. They print many casebound books for major publishers and book clubs (their advertisements say they can produce books in seven working days). They also offer fulfillment services.

We've been told by several smaller publishers that Haddon offers excellent prices on short runs of casebound books, but when we called Mr. Rossi he told us that they do not do short runs and are not interested in doing short runs.

Call one of their sales offices. Maybe you'll have better luck than we did.

User Comment: "On runs of 2000-5000 they are not rock bottom on prices but are great in service, quality and schedules."

RATINGS	1	2	3	4	5	6	7	8	9	10	Ave	
Speed	-	-	-	-	-	-	-	-	-	1	---	1
Price	-	-	-	-	-	-	1	-	-	-	---	
Dependability	-	-	-	-	-	-	-	-	-	1	---	
Service ...	-	-	-	-	-	-	-	-	-	1	---	
Quality ...	-	-	-	-	-	-	-	-	-	1	---	
Overall ...	-	-	-	-	-	-	-	-	-	1	---	

W. F. Hall Inc.   312-794-4600
4600 W Diversey Avenue   Gary W. McDearmon
Chicago IL 60639   Sales Representative

They have never answered any of our RFQ's or survey forms. They apparently do not do short runs.

```
* * * * * * * * * * * * * * * * * * * * * * * * * * * * * * *
*  Tip:  Remember:  The lowest price is not necessarily the    *
*        best bargain.  Don't sacrifice quality, service, or   *
*        delivery just to save a few dollars.                  *
* * * * * * * * * * * * * * * * * * * * * * * * * * * * * * *
```

Halliday Lithograph
Circuit Street
West Hanover MA 02339

617-826-8385 / 617-773-4697
John Baird
Vice President

Halliday is a division of the Arcata book group, which has a toll-free WATS line which you can call for quotes: 800-722-7020.

Quantities: Min: 500   Max: open   Opt: 3000 - 10,000

Book Sizes: 5 1/2 x 8 1/2; 6 x 9; 7 3/8 x 9 1/4; 8 1/2 x 11

Bindings: [X] PB   [X] SS   [X] HC   [X] C/SB

Capabilities: [ ] Magazines   [ ] Galley Copies
               [ ] Journals   [ ] Demand Printing
               [ ] Cookbooks   [ ] 4-color Juvenile Books
               [ ] Yearbooks   [ ] Annual Reports/Brochures
               [ ] Catalogs   [ ] Other Commercial Printing

Services: [ ] Typesetting   [ ] Teletypesetting
           [ ] Design and Pasteup [ ] Editing
           [ ] 4-color Printing   [ ] Warehousing/Shipping

Terms: Not stated.

Halliday, "the largest book manufacturing plant in the northeast," specializes in printing and binding high quality trade books and professional textbooks.

User Comment: "A heretofore persistent salesman has never called to follow up on the job or inquire after further work."

RATINGS	1	2	3	4	5	6	7	8	9	10	Ave	
Speed	1	-	-	-	-	-	-	-	-	-	---	1
Price	-	-	-	-	1	-	-	-	-	-	---	
Dependability	-	-	1	-	-	-	-	-	-	-	---	
Service	-	-	1	-	-	-	-	-	-	-	---	
Quality	-	-	-	-	-	-	-	1	-	-	---	
Overall	-	-	-	1	-	-	-	-	-	-	---	

Hamilton Printing Company
P O Box 232
Rensselaer NY 12144

518-477-9345
Brian Payne
President

A member of the Book Manufacturers Institute, they may be binding specialists. They did not answer our recent RFQ or printer survey.

Harlo Printing Company
50 Victor Avenue
Detroit MI 48203

313-883-3600   SP
Lothar W. Mueller
President

Quantities: Min: 250   Max: 20,000   Opt: 1000 - 20,000

Book Sizes: practically any size

Bindings: [X] PB   [ ] SS   [X] HC   [X] C/SB

Capabilities: [X] Magazines   [ ] Galley Copies
               [X] Journals   [ ] Demand Printing
               [X] Cookbooks   [X] 4-color Juvenile Books
               [ ] Yearbooks   [X] Annual Reports/Brochures
               [X] Catalogs   [X] Other Commercial Printing

Services: [X] Typesetting   [X] Teletypesetting
           [X] Design and Pasteup [X] Editing
           [X] 4-color Printing   [ ] Warehousing/Shipping

Terms: Depends on credit history; normally 50% down, balance
        prior to delivery.

Harlo has been in business since 1946. They specialize in
printing 5 1/2 x 8 1/2 books but can do other sizes as well.
They can also print magazines, newspapers, directories, and other
publications using either letterpress or offset lithography.

They have advertised in Writer's Digest for many years and,
therefore, have had much experience working with self-publishing
authors. They do quality work and are friendly, capable, and
easy to work with.

═══

Harmony Press
R.D. 2 / Box 123
Phillipsburg NJ 08865

?
Fred Grotenhuis
President

They did not answer the printer survey we sent them, but here
are some user comments regarding their services:

User Comments: "They are very good on small page counts, but
they're best at helping publishers with very little experience.
They can take a hand-written manuscript and return bound books.
They are very forgiving and probably offer the most complete
customer support of any printer I've dealt with. ... They have
made a speciality of very intricate collated material and can do
any projects that require a lot of handwork cheaper than the Ann
Arbor types or the big guns like Arcata or Donnelley."

RATINGS	1	2	3	4	5	6	7	8	9	10	Ave	
Speed	–	–	–	–	–	–	–	1	–	–	---	1
Price	–	–	1	–	–	–	–	–	–	–	---	
Dependability	–	–	–	–	–	–	–	1	–	–	---	
Service ...	–	–	–	–	–	–	–	–	–	1	---	
Quality ...	–	–	–	–	–	–	–	–	1	–	---	
Overall ...	–	–	–	–	–	–	–	1	–	–	---	

Hawkes Publishing Inc.   801-262-5555
3775 S Fifth W / P O Box 15711   John D. Hawkes
Salt Lake City UT 84115   President

They did not answer our recent printer survey, so we cannot detail their capabilities or services. Perhaps they are not capable of doing short runs.

Haymarket Press   612-721-4401
3451 Cedar Avenue S.   Nhat Hong
Minneapolis MN 55407   Attn: Marketing

Quantities: Min: 200   Max: 15,000   Opt: 3000 – 5000

Book Sizes: 5 1/2 x 8 1/2; 8 1/2 x 11

Bindings: [I] PB   [I] SS   [ ] HC   [I] C/SB

Capabilities: not stated

Services: none

Terms: 1/2 with order, 1/2 on delivery.

Haymarket is a cooperative, worker-run print shop. We have not seen their work, but they were well-recommended by a small Minnesota publisher. They did not answer our recent printer survey. It is possible that they are no longer doing short-run book printing.

*** Use this <u>Directory</u> to help you select the most likely printers to query for each book you want to publish.

Heart of the Lakes Publishing
2989 Lodi Road
Interlaken NY 14847-9763

607-532-4997
Walter Steesy
Owner

Quantities: Min: 250   Max: 10,000   Opt: 1000 - 3000

Book Sizes: 5 1/2 x 8 1/2; 6 x 9; 8 1/2 x 11

Bindings: [I] PB   [O] SS   [O] HC   [O] C/SB

Capabilities: [ ] Magazines   [ ] Galley Copies
               [X] Journals   [ ] Demand Printing
               [ ] Cookbooks   [ ] 4-color Juvenile Books
               [ ] Yearbooks   [ ] Annual Reports/Brochures
               [ ] Catalogs   [ ] Other Commercial Printing

Services: [X] Typesetting   [X] Teletypesetting
           [X] Design and Pasteup   [X] Editing
           [ ] 4-color Printing   [X] Warehousing/Shipping

Terms: Varies -- 1/3, 1/3, 1/3 pretty much standard.

They specialize in historical and genealogical publications.

User Comment: "Secured ISBN and copyright. Furnished advertising flyers and free warehousing. We believe this company to be a contractor or broker who is at the mercy of printers."

RATINGS	1	2	3	4	5	6	7	8	9	10	Ave	
Speed	–	–	–	1	–	–	–	–	–	–	---	1
Price	–	–	–	–	–	1	–	–	–	–	---	
Dependability	–	–	–	–	–	1	–	–	–	–	---	
Service . . .	–	–	–	–	–	–	–	–	1	–	---	
Quality . . .	–	–	–	–	–	–	–	1	–	–	---	
Overall . . .	–	–	–	–	–	–	–	1	–	–	---	

Heffernan Press
35 New Street
Worcester MA 01605

617-791-3661
William J. Heffernan II
President

They offer toll-free phone numbers for customers in Massachusetts (800-922-8229) and New England (800-343-6016).

They specialize in technical books, magazines, and journals as well as teachers manuals and catalogs. They have not answered our RFQ's or printer surveys, so we cannot give you any other details regarding their capabilities and services.

Henington Publishing Company
P O Drawer N
Wolfe City TX 75496

214-496-2226
Marc Wensel
Director of Manufacturing

Quantities: Min: 100   Max: 5000   Opt: 1000

Book Sizes: Any size between 5 1/2 x 8 1/2 and 9 x 12

Bindings: [I] PB   [I] SS   [I] HC   [I] C/SB

Capabilities: [ ] Magazines   [ ] Galley Copies
               [ ] Journals   [ ] Demand Printing
               [ ] Cookbooks   [ ] 4-color Juvenile Books
               [ ] Yearbooks   [ ] Annual Reports/Brochures
               [X] Catalogs   [X] Other Commercial Printing

Services: [X] Typesetting   [ ] Teletypesetting
           [ ] Design and Pasteup   [ ] Editing
           [X] 4-color Printing   [ ] Warehousing/Shipping

Terms: 1/3 down, 1/3 with proofs, balance on delivery.

A family-owned and operated business which does contract
printing for business and government as well as genealogies and
histories for self-publishing clients.

Heritage Printers
510 W Fourth Street
Charlotte NC 28202

704-372-5784
William E. Loftin
President

They specialize in hot metal composition and letterpress
printing of magazines and journals. They did not answer our
printer survey, so we cannot give any further details.

D. B. Hess Company
1150 McConnell Rd / PO Box 585
Woodstock IL 60098

815-338-6900
David B. Hess
President

They have sales offices in Massachusetts (617-664-3088) and
New Jersey (201-368-8999).

Quantities: Min: 5000   Max: 1,000,000 Opt: 10,000+

Book Sizes: 8 1/2 x 11 specialists

D. B. Hess Company continued

Bindings: [I] PB   [I] SS   [ ] HC   [I] C/SB

Capabilities: [ ] Magazines   [ ] Galley Copies
               [ ] Journals   [ ] Demand Printing
               [ ] Cookbooks   [ ] 4-color Juvenile Books
               [ ] Yearbooks   [X] Annual Reports/Brochures
               [X] Catalogs   [X] Other Commercial Printing

Services: [ ] Typesetting   [ ] Teletypesetting
           [ ] Design and Pasteup [ ] Editing
           [X] 4-color Printing   [X] Warehousing/Shipping

Terms: Net 30 with approved credit.

They have two divisions: educational (workbooks, lab manuals, and test booklets) and commercial (catalogs, annual reports). Both divisions specialize in one standard format: 8 1/2 x 11 books in one, two, or four colors. They offer their best prices on quantities of 10,000 or more. As a fairly new company (six years old), they are eager to work with you.

Hignell Printing Ltd.   204-783-7237
488 Burnell Street   Michael Ballnik
Winnepeg, Manitoba   Sales Manager
R3G 2B4   Canada

Quantities: Min: 250   Max: 10,000   Opt: 5000

Book Sizes: almost any size including some special sizes

Bindings: [I] PB   [I] SS   [I] HC   [X] C/SB

Capabilities: [X] Magazines   [ ] Galley Copies
               [X] Journals   [ ] Demand Printing
               [X] Cookbooks   [X] 4-color Juvenile Books
               [X] Yearbooks   [X] Annual Reports/Brochures
               [X] Catalogs   [ ] Other Commercial Printing

Services: [X] Typesetting   [ ] Teletypesetting
           [X] Design and Pasteup [ ] Editing
           [X] 4-color Printing   [ ] Warehousing/Shipping

Terms: Net 30 days.

Hignell is a Canadian printer specializing in short runs via either letterpress or offset.

A. B. Hirschfeld Press
5200 Smith Road
Denver CO 80216

303-320-8500
A. Barry Hirschfeld
President

Quantities: Min: 5000   Max: 250,000   Opt: 50,000

Book Sizes: 5 1/2 x 8 1/2; 6 x 9; 7 1/2 X 11; 8 1/2 x 11

Bindings: [I] PB   [I] SS   [O] HC   [O] C/SB

Capabilities: [X] Magazines   [X] Galley Copies
               [ ] Journals   [X] Demand Printing
               [X] Cookbooks   [ ] 4-color Juvenile Books
               [ ] Yearbooks   [X] Annual Reports/Brochures
               [X] Catalogs   [X] Other Commercial Printing

Services: [X] Typesetting   [X] Teletypesetting
           [X] Design and Pasteup   [ ] Editing
           [X] 4-color Printing   [X] Warehousing/Shipping

Terms: Discounts on approved credit.

Hirschfeld is a versatile commercial printer who also prints softcover books and magazines. They can also fulfill subscriptions for magazines and books (but cannot provide warehousing). They have been in business since 1907.

Holladay Tyler Printing Corp.
1900 Chapman Avenue
Rockville MD 20852

301-881-8050
E. Wayne Tyler
President

Holladay, a member of the Book Manufacturers Institute with over $40,000,000 in annual sales, is one of the 50 largest printers in the United States. However, they apparently do not do short runs since they have never answered any of our RFQ's or printer surveys.

```
* * * * * * * * * * * * * * * * * * * * * * * * * * * * * * * *
*  Tip:   You can save money by gang-running your books (that   *
*         is, by printing several books at the same time).  If  *
*         you can arrange to publish several books in the same  *
*         size and quantity, you can save on prep costs while   *
*         getting better quantity discounts on paper costs.     *
*         You will also save time and money in preparing manu-  *
*         scripts for typesetting and in designing your books   *
*         if you have a standard format for all your books.     *
* * * * * * * * * * * * * * * * * * * * * * * * * * * * * * * *
```

Hooven-Dayton Corporation
430 Leo Street
Dayton OH 45404

513-224-1108
Mrs. Thecla R. Zech
Vice President

Quantities: Min: 500   Max: 75,000   Opt: 25,000

Book Sizes: 5 1/2 x 8 1/2; 8 1/2 x 11

Bindings: [I] PB   [I] SS   [ ] HC   [I] C/SB

Capabilities: [X] Magazines   [ ] Galley Copies
                [X] Journals   [X] Demand Printing
                [X] Cookbooks   [X] 4-color Juvenile Books
                [ ] Yearbooks   [X] Annual Reports/Brochures
                [X] Catalogs   [X] Other Commercial Printing

Services: [X] Typesetting   [ ] Teletypesetting
            [ ] Design and Pasteup   [ ] Editing
            [ ] 4-color Printing   [X] Warehousing/Shipping

Terms: Net 30 days.

They have been printing for the U. S. Government Printing Office for over 30 years but are now branching out to do more private sector work. They specialize in printing one-color soft-cover books in the standard sizes.

Carl Hungness Publishing
P O Box 24308
Speedway IN 46224

317-244-4792
Terri Gunn
Sales Representative

They apparently do not do short runs since they did not answer our recent printer survey. They have both letterpress and offset printing facilities.

Hunter Publishing Company
2505 Empire Drive
Winston-Salem NC 27103

?
Raleigh Hunter III
Vice President of Marketing

They won the 1984 PIA Graphic Arts award for the printing of yearbooks. Since they did not answer our printer survey, we cannot give you any further details concerning their capabilities and services. They may be specialists in yearbooks only and not capable of doing other sorts of books.

Book Manufacturing

Jostens is a full service book manufacturer, serving the short-run production needs of publishers of trade books, professional books, religious books and university press.

Hard-cover, soft-cover and mechanically bound books are produced at each of the complete manufacturing facilities strategically located to reduce freight charges to distribution points.

Each plant is a marvel of self-contained efficiency, completely capable of providing for the short-run, rapid response requirements of today's book publishers. Electronic composition, color separation, 4 color presswork, perfect binding, smythe sewing and case binding all come together with precision timing to assure swift, accurate response to your market's demands.

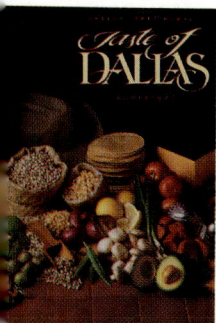

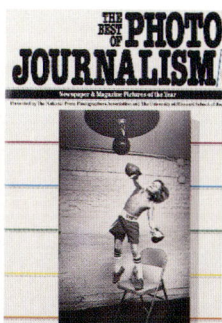

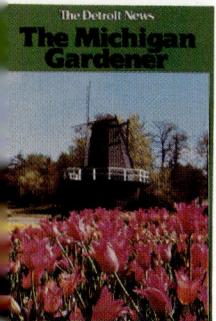

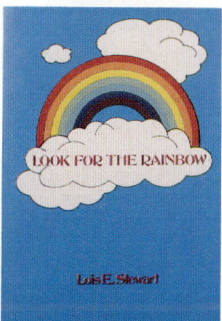

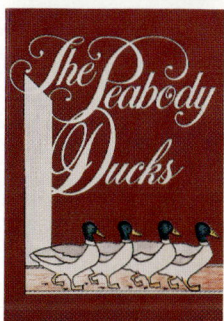

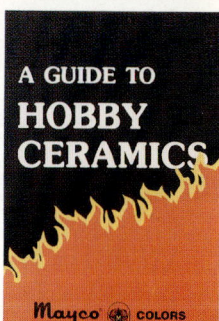

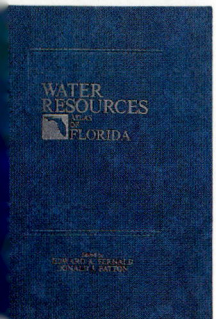

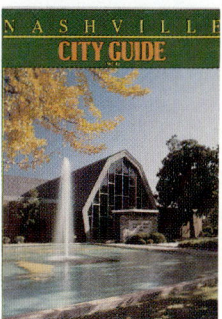

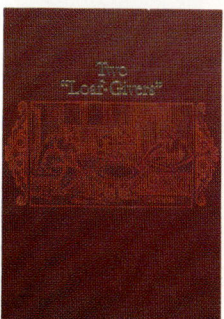

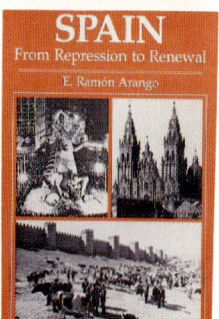

Topeka, Kansas

State College, Pennsylvania

Clarksville, Tennessee

Visalia, California

Complete 4 Color Printing & Binding Services.

Independence Press
3225 S Noland Road
Independence MO 64055

816-252-5010
Attn: Sales Representative

They are listed in the Kansas City phone book as being able to print books but they have never answered any of our RFQ's or survey forms. They apparently do not do short runs.

===

Independent Printing Company
141 East 25th Street
New York NY 10010

212-689-5100 / 914-949-9340
Norman Warwick
Sales Promotion

Quantities: Min: 25   Max: 2000   Opt: 1000

Book Sizes: 5 1/2 x 8 1/2; 6 x 9; 8 1/2 x 11

Bindings: [I] PB   [I] SS   [ ] HC   [I] C/SB

Capabilities:
- [ ] Magazines
- [X] Journals
- [ ] Cookbooks
- [ ] Yearbooks
- [ ] Catalogs

- [X] Galley Copies
- [X] Demand Printing
- [ ] 4-color Juvenile Books
- [X] Annual Reports/Brochures
- [X] Other Commercial Printing

Services:
- [X] Typesetting
- [ ] Design and Pasteup
- [ ] 4-color Printing

- [ ] Teletypesetting
- [ ] Editing
- [X] Warehousing/Shipping

Other services: blueprints, drafting and engineering supplies

Terms: Net 30 with credit approval; otherwise, 50% down, balance C.O.D.

They are specialists in printing and binding ultra-short runs -- anything from 25 copies up to 2500 copies. They also offer complete warehousing and fulfillment services.

```
* * * * * * * * * * * * * * * * * * * * * * * * * * * * * * * *
*  Tip:  It is important to accurately check specifications as  *
*        well as quoted prices when you receive quotes back     *
*        from printers, since many printers may change one or   *
*        two specifications to save costs or to fit your book   *
*        to their capabilities. The most common changes are     *
*        paper stock and trim size (for example, 5 3/8 x 8 3/8  *
*        trim size rather than 5 1/2 x 8 1/2). Be sure that     *
*        any such changes are acceptable to you.                *
* * * * * * * * * * * * * * * * * * * * * * * * * * * * * * * *
```

Inter-Collegiate Press
6015 Travis Lane / P O Box 10
Shawnee Mission KS 66201

913-432-8100
David Plumer
Manager

Quantities: Min: 100   Max: 10,000   Opt: 2000 - 5000

Book Sizes: any size from 4 1/2 x 6 to 9 x 12

Bindings: [I] PB   [I] SS   [I] HC   [I] C/SB

Capabilities: [X] Magazines   [ ] Galley Copies
                 [X] Journals   [ ] Demand Printing
                 [X] Cookbooks   [ ] 4-color Juvenile Books
                 [X] Yearbooks   [ ] Annual Reports/Brochures
                 [X] Catalogs   [ ] Other Commercial Printing

Services: [X] Typesetting   [X] Teletypesetting
             [ ] Design and Pasteup   [ ] Editing
             [X] 4-color Printing   [ ] Warehousing/Shipping

Terms: By arrangement, usually net 30 with approved credit.

Inter-Collegiate has been printing books since 1910. They
have a separate cover making plant that offers hot-stamped foil,
silkscreened, or embossed covers. They can print anything from
mass-market paperbacks to standard trade paperbacks and casebound
books to fancy yearbooks.

The samples of their work which we have seen are of excellent
quality with superb halftone reproduction. Inter-Collegiate was
rated among the top five book printers in the 1981 Small Press
Review printer survey. But note that 2 out of the 3 users who
rated them in our recent survey gave them only an average rating.

You can get a free gift from them if you mention their ad in
this edition of the Directory (on page 13) when you query them.

User Comment: "Super printers! Customer service is best in
business. Their prices are a little higher but worth it."

RATINGS	1	2	3	4	5	6	7	8	9	10	Ave	
Speed	-	-	-	-	1	-	1	1	-	-	6.67	3
Price	-	-	-	-	-	2	1	-	-	-	6.33	
Dependability	-	-	-	1	1	-	-	-	-	1	6.33	
Service . . .	-	-	-	-	1	1	-	-	-	1	7.0	
Quality . . .	-	-	-	1	1	-	-	-	-	1	6.33	
Overall . . .	-	-	-	1	1	-	-	-	-	1	6.33	

*** Use this Directory to help you select the most likely
    printers to query for each book you want to publish.

98

Interstate Book Manufacturers   800-255-0003 / 913-764-5600
2115 E Kansas City Road   Jim Axford
Olathe KS 66201   Estimator

In New York call 212-695-4505.

Quantities: Min: 1000   Max: open   Opt: 5000+

Book Sizes: any size from 3 x 5 to 9 x 12 or larger

Bindings: [I] PB   [I] SS   [I] HC   [I] C/SB

Capabilities: [ ] Magazines   [ ] Galley Copies
                [ ] Journals   [ ] Demand Printing
                [ ] Cookbooks   [X] 4-color Juvenile Books
                [ ] Yearbooks   [ ] Annual Reports/Brochures
                [X] Catalogs   [ ] Other Commercial Printing

Services: [X] Typesetting   [ ] Teletypesetting
            [ ] Design and Pasteup   [ ] Editing
            [ ] 4-color Printing   [X] Warehousing/Shipping

Terms: Net 30 days with approved credit.

   Interstate, now owned by Thomas Nelson Publishers, has just
added a 200,000 sq. ft. addition to their printing plant. They
specialize in 1 and 2-color books of any trim size from 3 x 5 to
9 x 12. They shy away from doing long runs in saddlestitched or
combbound books. They can print 4-color covers but not text.

===

Interstate Printers and Publrs   217-446-0500
19 N Jackson St / P O Box 594   Russell W. Zurlinden
Danville IL 61832   Vice President of Marketing

Quantities: Min: 300   Max: 100,000   Opt: 15,000 - 30,000

Book Sizes: 5 1/2 x 8 1/2; 6 x 9; 6 3/4 x 9 1/2; 8 1/2 x 11

Bindings: [I] PB   [I] SS   [I] HC   [I] C/SB

Capabilities: [X] Magazines   [ ] Galley Copies
                 [X] Journals   [ ] Demand Printing
                 [X] Cookbooks   [ ] 4-color Juvenile Books
                 [ ] Yearbooks   [X] Annual Reports/Brochures
                 [X] Catalogs   [X] Other Commercial Printing

   For more information on their services and terms, turn the
page.

Interstate Printers and Publishers continued

Services: [X] Typesetting   [ ] Teletypesetting
           [X] Design and Pasteup   [X] Editing
           [X] 4-color Printing   [X] Warehousing/Shipping

Terms: Net 30 with approved credit.

   We do not know much about this company, but here is what they
say about themselves: "Interstate is a service oriented book
printer offering excellent composition through completed and
shipped books. Our concerned and proficient personnel will help
plan and produce your book, allowing you more free time to
perform other functions." You might want to query them for your
next book.

Interstate Printing Company   402-341-8028
2002-22 N 16th Street   Stan Erickson
Omaha NE 68103   Sales Manager

   They offer both letterpress and offset composition and printing. However, they have never answered any of our RFQ's or survey forms. They apparently do not do short runs.

Jersey Printing Company   201-436-4200
111 Linnet St / P O Box 79   Steven Schnoll
Bayonne NJ 07002   Vice President Sales

   They also offer both letterpress and offset composition and
printing. But they, too, have never answered any of our RFQ's or
survey forms. Apparently they are not interested in doing short
runs.

The Job Shop   617-548-9600
P O Box 305   David Shephard
Woods Hole MA 02543   Proprietor

Quantities: Min: 100   Max: 2500   Opt: 1000

Book Sizes: 5 1/2 x 8 1/2; 6 x 9; 8 1/2 x 11

Bindings: [I] PB   [I] SS   [O] HC   [I] C/SB

Capabilities: [ ] Magazines   [ ] Galley Copies
               [ ] Journals   [X] Demand Printing
               [ ] Cookbooks   [ ] 4-color Juvenile Books
               [ ] Yearbooks   [X] Annual Reports/Brochures
               [X] Catalogs   [X] Other Commercial Printing

Services: [X] Typesetting   [X] Teletypesetting
           [X] Design and Pasteup [ ] Editing
           [ ] 4-color Printing   [ ] Warehousing/Shipping

Terms: 1/3 down, 1/3 with proofs, 1/3 on delivery.

Their specialty is the composition of scientific manuals, technical manuals, and bibliographies. They can also print short runs or broker complete book manufacturing services for longer runs.

John Henry Company   517-484-5403
200 N Cedar Street   Richard Wellman
Lansing MI 48933

Formerly known as Wellman Printing, this company does do short-run book printing but was too late to get a full listing.

Johnson & Hardin Company   513-271-8874
3600 Red Bank Road   Andrew M. Jamison
Cincinnati OH 45227   Sales Representative

They have never answered any of our RFQ's or survey forms. They apparently do not do short runs.

Johnson Graphics   815-747-6511
Frentress Lake Road   Harland E. Mapes
East Dubuque IL 61025   General Manager

Johnson Graphics is a commercial printer which can do short-run book printing. They haven't answered our last two printer surveys and may no longer be interested in doing short-run books.

Johnson Publishing Company   303-443-1576
1880 S 57th Ct / P O Box 990   Donald Caven
Boulder CO 80306   Sales Manager

They can print brochures, paperback books, and other publications. But they may not be interested in printing short runs
since they have never answered our RFQ's or printer surveys.

Jostens Printing & Publishing   612-830-8415
5501 Norman Center Drive   Larry Apeland
Minneapolis MN 55437   National Sales Manager

Jostens has four book printing plants (but all requests for
quotations should be sent to the address above):

1) 1312 Dickson Highway, P. O. Box 923, Clarksville, TN 37040
      Phone: 615-647-5211
2) 401 Science Park Road, P. O. Box 297, State College, PA 16801
      Phone: 814-237-5771
3) 4000 S. Adams Street, P. O. Box 1903, Topeka, KS 66609
      Phone: 913-266-3300
4) 29625 Road 84, P. O. Box 991, Visalia, CA 93279
      Phone: 209-651-3300

Quantities: Min: 100   Max: 100,000   Opt: 10,000

Book Sizes: They can handle almost any size.

Bindings: [I] PB   [I] SS   [I] HC   [I] C/SB

Capabilities: [X] Magazines   [ ] Galley Copies
               [X] Journals   [ ] Demand Printing
               [X] Cookbooks   [X] 4-color Juvenile Books
               [X] Yearbooks   [X] Annual Reports/Brochures
               [X] Catalogs   [X] Other Commercial Printing

Services: [X] Typesetting   [X] Teletypesetting
           [X] Design and Pasteup [ ] Editing
           [X] 4-color Printing   [ ] Warehousing/Shipping

Terms: Net 30 with approved credit.

Probably the largest sheet-fed only printer in the U.S., they
can provide design, typesetting, color separations, printing, and
binding. They also have a complete cover manufacturing plant for
screening, embossing, foil-stamping, or laminating all types of
covers. Since they print many school yearbooks, they are used to
dealing with new customers to produce high quality books.

Julin Printing
225 S Locust Street
Monticello IA 52310

319-465-3558
Ruth Julin
President

Quantities: Min: 2000   Max: 100,000   Opt: 10,000 - 30,000

Book Sizes: 5 1/2 x 8 1/2; 6 x 9; 8 1/2 x 11; 9 x 12

Bindings: [I] PB   [I] SS   [O] HC   [I] C/SB

Capabilities: [X] Magazines   [ ] Galley Copies
                [ ] Journals   [ ] Demand Printing
                [X] Cookbooks   [ ] 4-color Juvenile Books
                [ ] Yearbooks   [X] Annual Reports/Brochures
                [X] Catalogs   [X] Other Commercial Printing

Services: [ ] Typesetting   [ ] Teletypesetting
           [X] Design and Pasteup [X] Editing
           [X] 4-color Printing   [X] Warehousing/Shipping

Terms: Net 30 days.

Kansas City Press
2101 Kansas City Rd / PO 1240
Olathe KS 66061

800-821-5745
Ronald K. Evans
President

They have a separate division, Cookbook Publishers, which specializes in printing cookbooks for fundraisers. Query Don Stout,
their marketing director, at their toll-free number: 800-227-7282
or at 913-764-5900.

Quantities: Min: 200   Max: 50,000   Opt: 1000

Book Sizes: 5 1/2 x 8 1/2; 6 x 9

Bindings: [O] PB   [ ] SS   [ ] HC   [I] C/SB

Capabilities: [ ] Magazines   [ ] Galley Copies
                [ ] Journals   [ ] Demand Printing
                [X] Cookbooks   [ ] 4-color Juvenile Books
                [ ] Yearbooks   [ ] Annual Reports/Brochures
                [ ] Catalogs   [ ] Other Commercial Printing

Services: [ ] Typesetting   [ ] Teletypesetting
           [ ] Design and Pasteup [ ] Editing
           [X] 4-color Printing   [ ] Warehousing/Shipping

For more details, turn to the next page.

Kansas City Press continued

Terms: Net 30 days.

They specialize in printing and binding color spiral-bound books. Their division, Cookbook Publishers, prints standard format cookbooks for fundraising organizations. They may be the world's largest publisher of such personalized cookbooks. Send for their cookbook kit and price list. They also distribute such cookbooks through their affiliate, The Collection.

===

Kimberly Press   805-964-6469
5390 Overpass Rd / P O Box 399   Bill McNally
Goleta CA 93116   President

Quantities: Min: 500   Max: 5000   Opt: 2500

Book Sizes: 5 1/2 x 8 1/2; 6 x 9; 7 x 10; 8 1/2 x 11

Bindings: [I] PB   [I] SS   [I] HC.   [ ] C/SB

Capabilities: [X] Magazines   [X] Galley Copies
               [X] Journals   [ ] Demand Printing
               [ ] Cookbooks   [ ] 4-color Juvenile Books
               [ ] Yearbooks   [ ] Annual Reports/Brochures
               [ ] Catalogs   [ ] Other Commercial Printing

Services: [ ] Typesetting   [ ] Teletypesetting
           [ ] Design and Pasteup [ ] Editing
           [X] 4-color Printing   [ ] Warehousing/Shipping

Terms: Not stated.

They specialize in producing scholarly books and journals. They say, "We are serious local printers in a high overhead resort area. Our prices will never be as low as printers in Michigan. Many of Santa Barbara's 100+ publishers choose to have their printing done here."

```
* * * * * * * * * * * * * * * * * * * * * * * * * * * * * * *
*  Tip:  Book papers are different from offset grades -- they  *
*        are more opaque, made to a consistent bulk (which is  *
*        important for accurately fitting covers), and usually *
*        have less filler and stronger fiber (making them more *
*        flexible so they tend to lie flatter).  But they also *
*        tend to be more expensive.  You must make the choice. *
* * * * * * * * * * * * * * * * * * * * * * * * * * * * * * *
```

Kingsport Press
P O Box 711
Kingsport   TN 37662

615-246-7131
Donald S. Solt
Marketing Manager

Kingsport is a division of Arcata, which has a toll-free WATS line you may call for quotes: 800-722-7020.

Quantities: Min: 5000 Max: 1,000,000 Opt: 15,000 - 25,000

Book Sizes: any size including special and odd sizes

Bindings: [I] PB   [I] SS   [I] HC   [I] C/SB

Capabilities: [X] Magazines   [ ] Galley Copies
              [X] Journals   [ ] Demand Printing
              [X] Cookbooks   [X] 4-color Juvenile Books
              [X] Yearbooks   [X] Annual Reports/Brochures
              [X] Catalogs   [X] Other Commercial Printing

Services: [X] Typesetting   [ ] Teletypesetting
          [X] Design and Pasteup [ ] Editing
          [X] 4-color Printing   [X] Warehousing/Shipping

Terms: Net 30 days.

Kingsport, the largest book printing plant in the world, offers complete book manufacturing, starting from handwritten manuscripts through to completed books. They can print any sort of book, from mass-market paperbacks to fancy Bibles and yearbooks. They also have a distribution facility capable of providing any fulfillment services required by publishers.

Because Kingsport is so big, smaller publishers can sometimes get lost in the shuffle. Their bureaucracy may drive you crazy. They are noted for being slow (especially on smaller runs). Kingsport sees itself as a "full service, customer-oriented book manufacturer."

User Comments: "They can do anything, literally. We have only used them for complex jobs, and they have done a quality job. They have also taken a long time. Kingsport is geared to the Holts and Harpers of this world, and they take their sweet time in getting a job done for a little guy." ... "Slow (45 days), economical, average quality." ... "Arcata, specifically Kingsport Press, has been ghastly to work with. Their sales and customer service people are all polite as can be, but their organization is so big, so departmentalized and strewn with red tape, that they are just not equipped to handle problems when they arise. Our one experience with them is a continuing nightmare too long to relate here."

Turn the page to see the results of our ratings survey.

RATINGS	1	2	3	4	5	6	7	8	9	10	Ave	
Speed	1	-	-	2	-	-	-	-	-	-	3.0	3
Price	-	-	-	1	-	-	1	1	-	-	6.33	
Dependability	-	-	1	-	-	-	1	1	-	-	6.0	
Service . . .	1	1	-	-	1	-	-	-	-	-	2.67	
Quality . . .	-	1	-	-	1	-	1	-	-	-	4.67	
Overall . . .	-	-	-	1	2	-	-	-	-	-	4.67	

Kni Book Manufacturing   714-956-7300 / 415-726-3677 SF
1231 S State College Boulevard   Dorothy Finch
Anaheim CA 92806   Customer Service

Quantities: Min: 100   Max: 20,000   Opt: 5000

Book Sizes: 4 1/4 x 7; 5 1/2 x 8 1/2; 6 x 9; 8 1/2 x 7;
             8 1/2 x 11

Bindings: [I] PB   [I] SS   [O] HC   [I] C/SB

Capabilities: [ ] Magazines   [ ] Galley Copies
               [X] Journals   [X] Demand Printing
               [ ] Cookbooks   [ ] 4-color Juvenile Books
               [ ] Yearbooks   [ ] Annual Reports/Brochures
               [ ] Catalogs   [ ] Other Commercial Printing

Services: [X] Typesetting   [ ] Teletypesetting
           [ ] Design and Pasteup [ ] Editing
           [ ] 4-color Printing   [ ] Warehousing/Shipping

Terms: Various.

In business for over 15 years, Kni seems to be inconsistent in producing quality books. But they also seem to be committed to making good on any problems. Their greatest strength: fast turnarounds.

User Comments: "Their finished product is often stiff as if the paper is run wrong grain. Good turnaround time, though." ... "After their first job for me, I should have known better; but Delta couldn't meet the deadline and I threw reason and my money out the window." ... "Very congenial to work with -- very professional." ... "A problem on the first run (1/3 the run had bad covers, skewed pages and/or missing signatures) was corrected with a second printing -- free -- with an additional 150 copies for my trouble. It took some talking, but they came through."

RATINGS	1	2	3	4	5	6	7	8	9	10	Ave	
Speed	–	–	–	–	1	–	–	1	1	1	8.0	4
Price	–	–	–	–	1	–	1	1	–	1	7.5	
Dependability	1	–	–	–	–	–	1	1	–	1	6.5	
Service . . .	1	–	–	–	–	–	–	2	–	–	5.67	
Quality . . .	1	–	–	–	–	1	–	1	–	–	5.0	
Overall . . .	1	–	–	–	–	–	1	1	–	–	5.33	

C. J. Krehbiel Company
3962 Virginia Avenue
Cincinnati OH 45227

513-271-6035
Robert E. Everett
Sales Manager

They have sales offices in the following places:

516-783-1052 Bellmore NY
312-648-0025 Chicago IL
615-366-6813 Nashville TN

212-736-7606 New York NY
401-934-0020 North Scituate RI
216-751-2960 Shaker Heights OH

Quantities: Min: 2500   Max: 100,000   Opt: not stated

Book Sizes: 5 1/2 x 8 1/2; 6 1/2 x 8 1/4; 6 1/2 x 9 1/4; 7 x 10;
8 1/2 x 11; 9 x 12

Bindings: [I] PB   [I] SS   [I] HC   [I] C/SB

Capabilities: [ ] Magazines   [ ] Galley Copies
[ ] Journals   [ ] Demand Printing
[ ] Cookbooks   [X] 4-color Juvenile Books
[ ] Yearbooks   [ ] Annual Reports/Brochures
[X] Catalogs   [ ] Other Commercial Printing

Services: [ ] Typesetting   [ ] Teletypesetting
[ ] Design and Pasteup   [ ] Editing
[X] 4-color Printing   [ ] Warehousing/Shipping

Terms: Net 30 with approved credit.

They specialize in printing trade books and workbooks in three
sizes (8 1/2 x 11, 6 1/2 x 8 1/4, and 6 1/2 x 9 1/4) in one, two,
or four colors. They have a complete bindery in-house.

```
* * * * * * * * * * * * * * * * * * * * * * * * * * * * * * * * *
*  Tip:  Query at least five to ten printers on every book,      *
*          especially when the format differs from your standard  *
*          format or when you're varying the quantity printed.    *
* * * * * * * * * * * * * * * * * * * * * * * * * * * * * * * * *
```

W. A. Krueger   602-948-5650
7301 E Helm Drive   John D. Netsel
Scottsdale AZ 85260   Vice President of Marketing

They have two divisions that specialize in shorter runs: the New Berlin Division for short book runs (16555 W Rogers Drive, New Berlin, WI 53151; phone: 414-784-2000) and the Senatobia Division for magazines (121 Matthews Drive, Senatobia, MS 38668; phone: 601-562-5252). All queries should be directed to their main address in Scottsdale.

Quantities: Min: 5000   Max: 1,000,000 Opt: not stated

Book Sizes: almost any size

Bindings: [I] PB   [I] SS   [I] HC   [I] C/SB

Capabilities: [X] Magazines   [ ] Galley Copies
               [X] Journals   [ ] Demand Printing
               [ ] Cookbooks   [X] 4-color Juvenile Books
               [ ] Yearbooks   [X] Annual Reports/Brochures
               [X] Catalogs   [X] Other Commercial Printing

Services: [X] Typesetting   [X] Teletypesetting
           [ ] Design and Pasteup [ ] Editing
           [X] 4-color Printing   [X] Warehousing/Shipping

Terms: not stated

W. A. Krueger is one of the ten largest printers in the United States, with annual sales over $200,000,000. They print many of the major weekly magazines and fancy catalogs. They won the 1984 PIA Graphic Arts award for best juvenile picture book.

Their New Berlin, Wisconsin plant specializes in printing softcover and casebound books (especially elementary and high school textbooks, reference sets, and continuity series) in quantities of 5000 on up. Their Senatobia, Mississippi plant is designed to gang-run quality magazines (trade, regional, technical, and professional) in an economical standard format in quantities from 5000 to 75,000.

===

The Lane Press   802-863-5555
305 St Paul Street   Dan Drumheller
Burlington VT 05401   President

They may not be interested in doing short runs since they did not answer our recent printer survey.

Lanman Company
1520 Eckington Place NE
Washington DC 20002

202-269-5400
Attn: Sales Representative

Lanman won the 1984 PIA Graphic Arts award for the best trade book. Apparently, though, they do not do short runs. They also did not answer our recent printer survey.

Latham Process Corporation
200 Hudson Street
New York NY 10013

212-966-4500
Raymond Gombach
President

Latham prints catalogs and financial documents. They do not do short runs.

Lawson Graphics
3620 Lakeshore Boulevard W
Toronto, Ontario
M8W 1N9   Canada

416-251-3171
J. A. Federico
Sales Representative

They specialize in 4-color printing on coated stock. They did not answer our recent survey, so we cannot give any further details regarding their capabilities or services.

Hal Leighton Printing Company
P O Box 3952
Hollywood CA 91605

213-983-1105
Melvin Powers
President

SP

This company, recommended by Melvin Powers in his books on mail order selling and self-publishing, is simply Melvin's print brokerage sideline. You can get better prices -- and better lead times -- elsewhere.

```
* * * * * * * * * * * * * * * * * * * * * * * * * * * * * * * *
*  Tip:  Plan your printing and publishing schedules.  At the   *
*        very least, you should have a schedule of what books   *
*        you will be publishing in the coming year.  Planning   *
*        will save you from the natural disasters of rush jobs. *
* * * * * * * * * * * * * * * * * * * * * * * * * * * * * * * *
```

Lettermation
457 McGlincey Lane #3
Campbell CA 95008

408-559-6577
Oliver Cook
Owner

Quantities: Min: 100   Max: open   Opt: 1000+

Book Sizes: 5 1/2 x 8 1/2; 6 x 9; 8 1/2 x 11 (their specialty)

Bindings: [O] PB   [O] SS   [ ] HC   [O] C/SB

Capabilities: [ ] Magazines   [ ] Galley Copies
                [ ] Journals   [X] Demand Printing
                [ ] Cookbooks   [ ] 4-color Juvenile Books
                [ ] Yearbooks   [X] Annual Reports/Brochures
                [X] Catalogs   [X] Other Commercial Printing

Services: [ ] Typesetting   [ ] Teletypesetting
            [ ] Design and Pasteup [ ] Editing
            [ ] 4-color Printing   [X] Warehousing/Shipping

Terms: 2% 10, net 30 with approved credit.

Lettermation is a local printer. They are only capable of
side-stitching; all other bindings must be subcontracted.

Letters Inc.
300 S Topeka
Wichita KS 67202

?
Steven Anderson

They have not answered our two most recent RFQ's and printer
surveys. They may no longer do short-run book printing.

Lexington Press
7 Oakland Street
Lexington MA 02173

617-862-8900
R. F. Sacco
President

They are listed in LMP as being short-run book printers, but
they have never answered any of our RFQ's or survey forms.

```
* * * * * * * * * * * * * * * * * * * * * * * * * * * * * * *
*  Tip:  You can save money by buying your own paper for your  *
*        books (if you know what you are doing).               *
* * * * * * * * * * * * * * * * * * * * * * * * * * * * * * *
```

Liberty York Graphic Industries 516-481-8500 / 212-242-7979
171 Greenwich Street   Arthur Raskin
Hempstead NY 11550   President

Quantities: Min: 10   Max: 100,000   Opt: 15,000

Book Sizes: 5 1/2 x 8 1/2; 6 x 9; 8 1/2 x 11; 9 x 12; 11 x 17

Bindings: [I] PB   [I] SS   [O] HC   [I] C/SB

Capabilities: [ ] Magazines   [ ] Galley Copies
               [X] Journals   [ ] Demand Printing
               [ ] Cookbooks   [ ] 4-color Juvenile Books
               [ ] Yearbooks   [X] Annual Reports/Brochures
               [X] Catalogs   [X] Other Commercial Printing

Services: [X] Typesetting   [ ] Teletypesetting
           [ ] Design and Pasteup   [ ] Editing
           [ ] 4-color Printing   [ ] Warehousing/Shipping

Terms: Net 30 days after credit okay.

  A full service commercial printer, they are short-run, multi-page specialists (newsletters, financial statements, reports, and brochures).

Lithocolor Press   312-345-5530
9825 W Roosevelt Road   John Matheson
Westchester IL 60153   General Manager

Quantities: Min: 1500   Max: 100,000   Opt: 20,000

Book Sizes: 4 1/4 x 7; 5 1/2 x 8 1/2; 6 x 9; 7 x 9; 8 1/2 x 11

Bindings: [I] PB   [I] SS   [O] HC   [O] C/SB

Capabilities: [X] Magazines   [ ] Galley Copies
               [X] Journals   [ ] Demand Printing
               [ ] Cookbooks   [ ] 4-color Juvenile Books
               [ ] Yearbooks   [ ] Annual Reports/Brochures
               [X] Catalogs   [X] Other Commercial Printing

Services: [ ] Typesetting   [ ] Teletypesetting
           [ ] Design and Pasteup   [ ] Editing
           [X] 4-color Printing   [X] Warehousing/Shipping

  For more about their terms and their specialties, turn the page.

Lithocolor Press continued

Other Services: fulfillment for magazines

Terms: Net 30 with approved credit.

They specialize in printing 6 x 9 and 7 x 9 textbooks and mass-market paperbacks (in quantities from 3000 to 50,000). They do many books for local Christian publishers.

Little River Press
55 NE 73rd Street
Miami FL 33138

305-757-7504
Richard Nettina
President

Quantities: Min: 1000   Max: 500,000   Opt: 15,000

Book Sizes: 4 1/4 x 7; 5 1/2 x 8 1/2; 8 1/2 x 11

Bindings: [I] PB   [I] SS   [O] HC   [ ] C/SB

Capabilities: [X] Magazines   [ ] Galley Copies
                [X] Journals   [ ] Demand Printing
                [ ] Cookbooks   [X] 4-color Juvenile Books
                [ ] Yearbooks   [X] Annual Reports/Brochures
                [X] Catalogs   [X] Other Commercial Printing

Services: [ ] Typesetting   [ ] Teletypesetting
            [ ] Design and Pasteup [ ] Editing
            [X] 4-color Printing   [ ] Warehousing/Shipping

Terms: 1% 10, net 30 days.

They were recommended to us by a publisher from Louisiana who says they offer excellent prices (apparently among the lowest in the country) and fast delivery. If you find that their price is right, be sure to see samples of their work before buying.

Lorain Book Manufacturers
2543 Broadway
Lorain OH 44052

216-244-3839
John Gallacher
President

SP

We received direct mail ads from this new printer, but they did not answer our RFQ or printer survey. Are they still in business?

Lorell Press
Bodwell St, Avon Industrial Pk
Avon MA 02322

617-471-7750
Joseph Scibeck
Sales Representative

They have never answered any of our RFQ's or survey forms.
They apparently do not do short runs.

Lorrah & Hitchcock Publishers
301 S 15th Street
Murray KY 42071

502-753-3759
Jean Lorrah
President

SP

Quantities: Min: 100   Max: 2000   Opt: 100 - 1000

Book Sizes: 5 1/2 x 8 1/2; 8 1/2 x 11

Bindings: [I] PB   [I] SS   [O] HC   [ ] C/SB

Capabilities: [X] Magazines   [X] Galley Copies
                 [X] Journals   [ ] Demand Printing
                 [X] Cookbooks   [ ] 4-color Juvenile Books
                 [ ] Yearbooks   [ ] Annual Reports/Brochures
                 [ ] Catalogs   [ ] Other Commercial Printing

Services: [X] Typesetting   [ ] Teletypesetting
             [X] Design and Pasteup   [X] Editing
             [ ] 4-color Printing   [ ] Warehousing/Shipping

Terms: Payment with order; check, money order, or credit card.

They serve self-publishing authors and handle everything from
handwritten manuscripts to camera-ready copy. Since they collate
by hand, they prefer doing runs of 100 to 500 copies. Send for
their free sample book, estimate forms, and price list.

```
* * * * * * * * * * * * * * * * * * * * * * * * * * * * * * * * *
*  Tip:  Get everything in writing.  Any modifications in your   *
*        specifications that you agree to over the phone or in   *
*        a conversation should be put into writing (either       *
*        included in the contract or attached thereto).  If      *
*        you don't require a signed contract when you work       *
*        with a printer, you should at least send a letter of    *
*        confirmation with your manuscript or camera-ready       *
*        copy.  This letter should reconfirm the specifica-      *
*        tions in the written quote you received from the        *
*        printer (including final price and delivery date).      *
*        Attach a copy of your original RFQ as well.             *
* * * * * * * * * * * * * * * * * * * * * * * * * * * * * * * * *
```

John D. Lucas Printing Company 301-633-4200
1820 Portal Street   William Krieger
Baltimore MD 21224   Vice President Sales

In Washington, DC call 202-621-4400; in New York call 212-947-6006; if you are in zone 1, you can call their toll-free number: 800-638-2850.

Quantities: Min: 1000   Max: 500,000   Opt: not stated

Book Sizes: almost any size

Bindings: [X] PB   [X] SS   [X] HC   [ ] C/SB

Capabilities: [X] Magazines   [ ] Galley Copies
                [X] Journals   [ ] Demand Printing
                [ ] Cookbooks   [ ] 4-color Juvenile Books
                [ ] Yearbooks   [X] Annual Reports/Brochures
                [X] Catalogs   [X] Other Commercial Printing

Services: [X] Typesetting   [ ] Teletypesetting
            [ ] Design and Pasteup [ ] Editing
            [X] 4-color Printing   [ ] Warehousing/Shipping

Terms: 1/3 down, 1/3 with proofs, 1/3 before shipment.

They print one, two, or four-color books, journals, catalogs, calendars, greeting cards, posters, and promotional material. They are most competitive at higher quantities.

User Comment: "Good people, fast service."

RATINGS	1	2	3	4	5	6	7	8	9	10	Ave
Speed	-	-	-	-	-	-	-	-	1	-	---	1
Price	-	-	-	-	-	-	1	-	-	-	---
Dependability	-	-	-	-	-	-	-	-	-	1	---
Service . . .	-	-	-	-	-	-	-	-	-	1	---
Quality . . .	-	-	-	-	-	-	-	-	-	1	---
Overall . . .	-	-	-	-	-	-	-	-	1	-	---

Mack Printing   215-258-9111
20th & Northampton Streets   Robert C. J. Miller
Easton PA 18042   Vice President of Marketing

Quantities: Min: 5000   Max: 150,000   Opt: 80,000

Book Sizes: 8 1/2 x 11 and tabloid size

Bindings: [I] PB   [I] SS   [ ] HC   [ ] C/SB

Capabilities: [X] Magazines   [ ] Galley Copies
                [X] Journals   [ ] Demand Printing
                [ ] Cookbooks   [ ] 4-color Juvenile Books
                [ ] Yearbooks   [X] Annual Reports/Brochures
                [X] Catalogs   [X] Other Commercial Printing

Services: [X] Typesetting   [X] Teletypesetting
            [X] Design and Pasteup   [X] Editing
            [X] 4-color Printing   [X] Warehousing/Shipping

Other services: database management

Terms: Net 30.

They specialize in short to medium-run scientific, technical, and trade publications.

Peter F. Mallon Inc.   212-786-2000
45-29 Court Square   Harry Mallon
Long Island City NY 11101   President

They do short runs, but they apparently have only letterpress facilities. They have never answered any of our RFQ's or survey forms.

Malloy Lithographing   800-722-3231 / 313-665-6113
5411 Jackson Rd / P O Box 1124   D. D. Robbins
Ann Arbor MI 48106   Customer Service

They have sales offices on both coasts:

212-662-2276 New York NY   201-447-2281 Waldwick NJ
415-482-1166 Oakland CA   312-668-7880 Wheaton IL

Quantities: Min: 500   Max: 100,000   Opt: 5000

Book Sizes: almost any size between 5 1/2 x 8 1/2 and 8 1/2 x 11

Bindings: [I] PB   [I] SS   [O] HC   [O] C/SB

For more details on their capabilities and services, turn the page.

Capabilities: [ ] Magazines   [ ] Galley Copies
                [X] Journals   [ ] Demand Printing
                [ ] Cookbooks   [ ] 4-color Juvenile Books
                [ ] Yearbooks   [ ] Annual Reports/Brochures
                [ ] Catalogs   [ ] Other Commercial Printing

Services: [ ] Typesetting   [ ] Teletypesetting
            [ ] Design and Pasteup   [ ] Editing
            [ ] 4-color Printing   [ ] Warehousing/Shipping

Terms: Net 30 with approved credit.

They have web capability for single color work in most sizes and sheet-fed capability for one or two colors; they can also do 4-color covers and inserts.

Malloy does a lot of work for smaller publishers and university presses. Their quality and service are both excellent. They were well-rated in the 1981 Small Press Review survey.

User Comments: "Turnaround: 30 days. Good price." ... "Good. Interested in clients. No longer as cheap as before." ... "Will work with you on special projects."

RATINGS	1	2	3	4	5	6	7	8	9	10	Ave	
Speed	–	–	–	1	–	2	–	–	–	–	5.33	3
Price	–	–	–	–	1	–	–	1	1	–	7.33	
Dependability	–	–	–	–	1	–	–	–	1	1	8.0	
Service . . .	–	–	–	–	–	1	–	1	1	–	7.67	
Quality . . .	–	–	–	–	–	1	–	–	1	1	8.33	
Overall . . .	–	–	–	–	–	1	1	–	1	–	7.33	

Maple-Vail Book Manufacturing   717-764-5911
Willow Springs Lane, P O 2695   Sam W. Vail
York PA 17405   Vice President Sales

They have sales offices in the following locations:

312-799-7510 Homewood IL   215-925-5950 Philadelphia PA
212-481-9150 New York NY   301-821-6690 Towson MD
617-965-1120 Newton MA   415-934-1440 Walnut Creek CA

They have another manufacturing plant, Vail-Ballou, at 187 Clinton Street, Binghamton, NY 13902; phone: 607-723-7981.

For details on their capabilities and services, see the page following the Malloy ad.

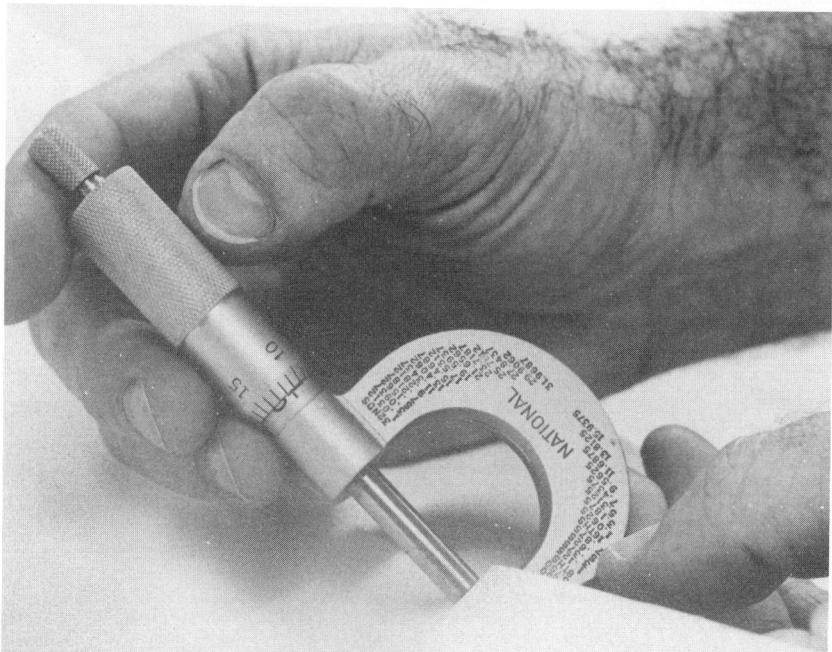

Quantities: Min: 500   Max: 250,000   Opt: not stated

Book Sizes: 5 1/2 x 8 1/2; 6 x 9; 8 1/2 x 11

Bindings: [I] PB   [ ] SS   [I] HC   [I] C/SB

Capabilities: [ ] Magazines   [ ] Galley Copies
               [ ] Journals   [ ] Demand Printing
               [X] Cookbooks   [ ] 4-color Juvenile Books
               [ ] Yearbooks   [ ] Annual Reports/Brochures
               [X] Catalogs   [ ] Other Commercial Printing

Services: [X] Typesetting   [ ] Teletypesetting
           [ ] Design and Pasteup   [ ] Editing
           [ ] 4-color Printing   [X] Warehousing/Shipping

Terms: To be established.

Maple-Vail, a member of the Book Manufacturers Institute, is one of the 50 largest printers in the United States, with annual sales of $45,000,000.

User Comments: "Very good quality, larger selection of paper, helpful, slower scheduling. Will invoice a paper clip if they supply you with one; slightly higher prices than we prefer." ... "Their prices seem to be awful. They were more than 30% higher than any other quote, over a range of 12 to 15 quotes covering both case and paper bound, various sizes and quantities." ... "They are our principal supplier. We consider them our best source based on the combination of price-quality-service." ... (Vail-Ballou): "Seem to have good prices on their big web, but they have lost art, failed to meet schedules, etc."

RATINGS	1	2	3	4	5	6	7	8	9	10	Ave	
Speed	-	-	-	1	-	1	-	-	-	-	---	2
Price	-	-	-	-	1	1	-	-	-	-	---	
Dependability	-	-	-	-	-	1	-	-	1	-	---	
Service . . .	-	-	-	-	-	1	-	-	1	-	---	
Quality . . .	-	-	-	-	-	-	1	-	1	-	---	
Overall . . .	-	-	-	-	-	-	1	-	1	-	---	

RATINGS -- Vail Ballou	3	4	5	6	7	8	9	10	Ave	
Speed	-	-	1	-	-	-	-	-	-	---	1
Price	-	-	-	-	1	-	-	-	-	---	
Dependability	-	-	1	-	-	-	-	-	-	---	
Service . . .	-	-	1	-	-	-	-	-	-	---	
Quality . . .	-	-	-	-	1	-	-	-	-	---	
Overall . . .	-	-	-	-	1	-	-	-	-	---	

Marek Lithographics Inc.   414-679-3600
S81 W18460 Gemini Dr / P 0 904   Pam Mallas
Muskego WI 53150   Inside Sales

They asked to be listed in the <u>Directory</u> but were too late to
get a full listing.

===

Mark IV Press Ltd.   516-349-8070
45 Fairchild Avenue   Robert Rinere
Plainview NY 11803   President

They also asked to be listed in the <u>Directory</u> but were too
late to get a full listing.

===

Mast Printing   ?
RR 4 / Box 130-A   Attn: President
Millersburg OH 44654

They did not answer the printer survey form we sent them after
we learned about them from one of their customers (see rating
below). Hence, we cannot give any details regarding their capabilities or services.

User Comment: "Very accomodating."

RATINGS	1	2	3	4	5	6	7	8	9	10	Ave	
Speed	-	-	-	-	-	-	1	-	-	-	---	1
Price	-	-	-	-	-	-	-	-	1	-	---	
Dependability	-	-	-	-	-	-	-	1	-	-	---	
Service . . .	-	-	-	-	-	-	-	1	-	-	---	
Quality . . .	-	-	1	-	-	-	-	-	-	-	---	
Overall . . .	-	-	-	-	-	-	1	-	-	-	---	

```
* * * * * * * * * * * * * * * * * * * * * * * * * * * * * * * *
*  Tip:  Never ask for delivery ASAP; always specify the exact  *
*        date when you expect delivery of the completed job.    *
*        In turn, always stand by your own commitments to get   *
*        camera-ready materials to the printer on time.  If     *
*        something holds you up, be sure to let the printer     *
*        know in plenty of time.  Otherwise, you might be       *
*        charged for the printer's downtime.                    *
* * * * * * * * * * * * * * * * * * * * * * * * * * * * * * * *
```

Maverick Publications
P O Drawer 5007
Bend OR 97708

503-382-6978
Gary Asher
Manager

Quantities: Min: 200   Max: 10,000   Opt: 1000 - 2000

Book Sizes: 5 1/2 x 8 1/2; 6 x 9; 8 1/2 x 11

Bindings: [I] PB   [I] SS   [O] HC   [I] C/SB

Capabilities: [ ] Magazines   [ ] Galley Copies
               [X] Journals   [ ] Demand Printing
               [X] Cookbooks   [ ] 4-color Juvenile Books
               [ ] Yearbooks   [ ] Annual Reports/Brochures
               [X] Catalogs   [X] Other Commercial Printing

Services: [X] Typesetting   [X] Teletypesetting
           [X] Design and Pasteup [X] Editing
           [X] 4-color Printing   [ ] Warehousing/Shipping

Terms: 1/2 with order, 1/2 on approval of page proofs.

   They were one of the first book printers to specialize in
serving self-publishers. They also print direct mail packages
and computer manuals. The work we have seen has always looked
good, but we have heard some questions about how well their bindings hold up. Write or call for their printed rate sheet and a
free copy of their book, How to Publish Your Book and Make It a
Best Seller.

   User Comment: "Excellent quality of product compensated for
production delay."

RATINGS	1	2	3	4	5	6	7	8	9	10	Ave
Speed	-	-	1	-	-	-	-	-	-	-	---	1
Price	-	-	-	-	1	-	-	-	-	-	---
Dependability	-	-	-	1	-	-	-	-	-	-	---
Service . . .	-	-	-	-	1	-	-	-	-	-	---
Quality . . .	-	-	-	-	-	-	-	-	-	1	---
Overall . . .	-	-	-	-	-	1	-	-	-	-	---

```
* * * * * * * * * * * * * * * * * * * * * * * * * * * * * * * *
*  Tip:  Provide your printer with all the camera-ready copy,  *
*        photographs, inserts, and instructions at the same    *
*        time.  There is far less room for mistakes to creep   *
*        into the job if the printer gets everything at once.  *
*        If you have stayed with your schedule, you should     *
*        have no trouble getting everything together to be     *
*        sent at the same time.                                *
* * * * * * * * * * * * * * * * * * * * * * * * * * * * * * * *
```

McClain Printing Company   304-478-2881   SP
212 Main Street   George A. Smith Jr.
Parson WV 26287   President

Quantities: Min: 250   Max: 10,000   Opt: 1000

Book Sizes: 5 1/2 x 8 1/2; 6 x 9; 8 1/2 x 11

Bindings: [I] PB   [I] SS   [O] HC   [I] C/SB

Capabilities: [X] Magazines   [X] Galley Copies
               [X] Journals   [ ] Demand Printing
               [ ] Cookbooks   [ ] 4-color Juvenile Books
               [ ] Yearbooks   [X] Annual Reports/Brochures
               [ ] Catalogs   [X] Other Commercial Printing

Services: [X] Typesetting   [X] Teletypesetting
           [X] Design and Pasteup [ ] Editing
           [X] 4-color Printing   [X] Warehousing/Shipping

Terms: Net 30 days.

   Besides doing commercial printing for the local area, McClain
offers book production and distribution services to authors who
want to self-publish. Every book they print is listed free in
"McClain Imprints," a book catalog that is sent to over 12,500
bookstores and libraries across the country. McClain will also
assign an ISBN number, obtain an LC number, and copyright books
for self-publishers.

===

McDowell Publications   ?
Rt 4 / Box 314   Attn: President
Utica KY 42376

   They did not answer our printer survey form; they may no
longer be in business.

   User Comment: "They may have upgraded their plant. If not,
I do not recommend them."

RATINGS	1	2	3	4	5	6	7	8	9	10	Ave	
Speed	-	-	-	-	-	-	1	-	-	-	---	1
Price	-	-	-	1	-	-	-	-	-	-	---	
Dependability	-	-	-	-	1	-	-	-	-	-	---	
Service	-	-	-	-	-	-	1	-	-	-	---	
Quality	-	1	-	-	-	-	-	-	-	-	---	
Overall	-	-	-	1	-	-	-	-	-	-	---	

The McFarland Company
P O Box 3645
Harrisburg PA 17105

717-234-6235
Thomas Fitzgerald
Vice President

They are listed in LMP as doing short runs, but they have never answered any of our RFQ's or survey forms.

McGregor & Werner
6411 Chillum Place NW
Washington DC 20012

202-722-2200
David B. Dobson
Managing Director

They have two printing plants: St. Mary's Press, Airport View Drive, Hollywood, MD 20636 (phone: 301-373-5827) and Graphic Printing Division, 310 N. Clay Street, New Carlisle, OH 45344 (phone: 513-845-3752).

Quantities: Min: 100   Max: 10,000   Opt: 2000 - 5000

Book Sizes: almost any size, including oversized and odd sizes

Bindings: [I] PB   [I] SS   [O] HC   [I] C/SB

Capabilities: [ ] Magazines   [X] Galley Copies
               [X] Journals   [X] Demand Printing
               [X] Cookbooks   [ ] 4-color Juvenile Books
               [ ] Yearbooks   [ ] Annual Reports/Brochures
               [ ] Catalogs   [X] Other Commercial Printing

Services: [ ] Typesetting   [ ] Teletypesetting
           [ ] Design and Pasteup [ ] Editing
           [ ] 4-color Printing   [ ] Warehousing/Shipping

Terms: Depends entirely on creditworthyness -- each case is unique. No prejudgments either way!

They specialize in conference publications and reprints; they can print and ship in 10 working days (or less, with planning), but their normal turnaround time is 15 working days. They also print newsletters, directories, business forms, computer forms, programs, posters, and, of course, books and journals. They can do both softcover and hardcover books in quantities as low as 100 copies at an economical price.

User Comments: "I was satisfied with the work and speed with which it was done; the quality of the printing was not outstanding, however their prices are extremely competitive."... "Good."

See their ratings chart on the page following their ad.

McGregor & Werner continued

RATINGS	1	2	3	4	5	6	7	8	9	10	Ave
Speed	-	-	-	-	-	-	-	2	-	-	---	2
Price	-	-	-	-	-	-	1	-	1	-	---
Dependability	-	-	-	-	-	1	-	1	-	-	---
Service . . .	-	-	-	-	-	1	1	-	-	-	---
Quality . . .	-	-	-	-	1	-	-	-	1	-	---
Overall . . .	-	-	-	-	-	-	1	1	-	-	---

McNaughton & Gunn
960 Woodland Dr / P O Box 128
Saline MI 48176

313-429-5411
Ronald A. Mazzola
Marketing Director

They also have an Ann Arbor address: P. O. Box M 2060, Ann Arbor, MI 48106. They have sales offices in the following areas:

312-537-0169 Chicago IL
212-486-7850 New York NY

415-927-0111 San Francisco CA
904-893-5563 Tallahassee FL

Quantities: Min: 50   Max: 20,000   Opt: 5000

Book Sizes: almost any size including odd sizes

Bindings: [I] PB   [I] SS   [O] HC   [I] C/SB

Capabilities: [ ] Magazines   [ ] Galley Copies
                [X] Journals   [ ] Demand Printing
                [X] Cookbooks   [X] 4-color Juvenile Books
                [ ] Yearbooks   [ ] Annual Reports/Brochures
                [X] Catalogs   [ ] Other Commercial Printing

Services: [ ] Typesetting   [ ] Teletypesetting
            [ ] Design and Pasteup   [ ] Editing
            [ ] 4-color Printing   [ ] Warehousing/Shipping

Terms: Net 30 with approved credit.

Founded in 1975, McNaughton & Gunn has already established themselves as one of the top short-run book printers in the country. They tied for first in the 1981 Small Press Review survey. Nonetheless, as you will see in the comments and ratings on the next page, they are somewhat inconsistent. For the most part, they get high ratings for quality of work, but more variable ratings for service and dependability. Why the inconsistency? We don't know. Perhaps it is growing pains. Be sure to make it clear to them exactly what your expectations are when you order.

If you want odd-sized books, try M&G first. They are odd-size specialists. Send for their unusual book brochure and sample copy. They can do beautiful things with odd sizes.

User Comments: "Most consistent in quality and good prices. Turnaround time average and wish it were better." ... "Very un-satisfactory ... more problems than all our other manufacturing combined."... "Careless about handling details." ... "Excellent in all respects." ... "Good work, good price, deliver when they promise." ... "Very professional and easy to work with in all respects. The quality of their work is exceptional." ... "Big problems with their shipper. Will not use again." ... "Average quality, hassle-free, friendly, refuse to negotiate prices, slightly inferior lay-up quality."

RATINGS	1	2	3	4	5	6	7	8	9	10	Ave	
Speed	1	–	–	–	5	1	3	1	1	–	5.83	12
Price	–	–	–	1	1	1	–	–	4	5	8.42	
Dependability	2	–	–	–	2	2	1	–	2	3	6.58	
Service . . .	1	1	–	–	2	–	2	2	2	2	6.75	
Quality . . .	–	–	–	1	3	1	–	1	2	4	7.58	
Overall . . .	–	2	–	–	3	–	2	–	2	3	6.75	

Meaker the Printer
802 W Jefferson Street
Phoenix AZ 85007

602-254-2171
Thom Meaker
President

Quantities: Min: 250   Max: open   Opt: 5000 – 10,000

Book Sizes: almost any size

Bindings: [O] PB   [I] SS   [O] HC   [I] C/SB

Capabilities: [X] Magazines   [X] Galley Copies
                [X] Journals   [X] Demand Printing
                [X] Cookbooks   [ ] 4-color Juvenile Books
                [ ] Yearbooks   [X] Annual Reports/Brochures
                [X] Catalogs   [X] Other Commercial Printing

Services: [X] Typesetting   [X] Teletypesetting
            [X] Design and Pasteup [X] Editing
            [X] 4-color Printing   [ ] Warehousing/Shipping

Terms: Net 10 days with approved credit.

General commercial printers who can also do books, they emphasize their customer service.

Meriden-Stinehour
47 Billard Street
Meriden CT 06450

203-235-7929 / 212-410-1455
Stephen Stinehour
Vice President

They also have another division Stinehour Press in Lunenberg, VT 05906; phone: 802-328-2507.

Quantities: Min: 10   Max: 100,000   Opt: 3000 - 6000

Book Sizes: 5 1/2 x 8 1/2; 6 x 9; 8 1/2 x 11; and oversized

Bindings: [I] PB   [I] SS   [I] HC   [X] C/SB

Capabilities: [X] Magazines   [X] Galley Copies
                [X] Journals   [X] Demand Printing
                [X] Cookbooks   [X] 4-color Juvenile Books
                [ ] Yearbooks   [X] Annual Reports/Brochures
                [X] Catalogs   [X] Other Commercial Printing

Services: [X] Typesetting   [X] Teletypesetting
            [X] Design and Pasteup [X] Editing
            [X] 4-color Printing   [X] Warehousing/Shipping

Terms: Net 15 days.

Stinehour Press specializes in short-run scholarly publications and fine limited editions. Meriden specializes in high quality reproduction of illustrated books. They offer both letterpress and offset composition and printing with full design and editing capabilities. They also offer mailing services to individual subscribers for journals and other periodicals. They can also arrange warehousing and drop shipping for books. They won several 1984 AIGA awards for graphic excellence.

===

Messenger Graphics
1207 E Washington / Box 29096
Phoenix AZ 85030

602-254-7231
Robert Hughes
General Manager

They specialize in color printing and can do short runs, but they have never answered our RFQ's or printer surveys. Perhaps they only want to work with local publishers.

```
* * * * * * * * * * * * * * * * * * * * * * * * * * * * * * * *
*  Tip:  Try to figure out your true costs for producing books  *
*        -- including the cost of your time.  Value your time.  *
*        It may be your major expense in producing books.       *
* * * * * * * * * * * * * * * * * * * * * * * * * * * * * * * *
```

Metromail
901 W Bond
Lincoln NE 68521

402-475-4591
Joe Rowland
Marketing Manager

Quantities: Min: 200   Max: 5000   Opt: 1500

Book Sizes: 8 1/2 x 11

Bindings: [I] PB   [I] SS   [O] HC   [O] C/SB

Capabilities: [ ] Magazines   [ ] Galley Copies
               [ ] Journals   [ ] Demand Printing
               [ ] Cookbooks   [ ] 4-color Juvenile Books
               [ ] Yearbooks   [ ] Annual Reports/Brochures
               [ ] Catalogs   [ ] Other Commercial Printing

Services: [X] Typesetting   [ ] Teletypesetting
           [ ] Design and Pasteup   [ ] Editing
           [ ] 4-color Printing   [ ] Warehousing/Shipping

Terms: Not stated.

   Metromail specializes in short-run printing of large-size
8 1/2 x 11) directories and other high page count books, typeset
from mag tape or floppy discs. Indeed, as you can see from their
listing above, they don't do much else.

===

Micro Book Manufacturing Co.
723 S Wells Street
Chicago IL 60607

312-922-2083
Kendall Victorine
Sales Manager

   They specialize in short-run 4-color books, inserts, covers,
and brochures. They have never answered our printer surveys, so
we cannot give you any other details regarding their services.

===

Mitchell Press Ltd.
1706 W First Avenue, Box 6000
Vancouver, British Columbia
V6B 4B9   Canada

604-731-5211
Jack Mellor
Sales Manager

   These Canadian printers have typesetting, design, and 4-color
printing capabilities. They can also do saddlestitching and perfectbinding. Other than that, we don't know too much about them.
They did not answer our recent printer survey.

Mitchell-Shear
713 W Ellsworth Road
Ann Arbor MI 48104

313-995-2505
Wayne A. Johnson
Operations Manager

Quantities: Min: 200   Max: depends   Opt: 3000 - 6000

Book Sizes: 5 1/2 x 8 1/2; 6 x 9; 8 1/2 x 11

Bindings: [I] PB   [I] SS   [O] HC   [I] C/SB

Capabilities: [X] Magazines   [X] Galley Copies
               [X] Journals   [X] Demand Printing
               [X] Cookbooks   [X] 4-color Juvenile Books
               [ ] Yearbooks   [X] Annual Reports/Brochures
               [X] Catalogs   [X] Other Commercial Printing

Services: [X] Typesetting   [ ] Teletypesetting
           [X] Design and Pasteup   [ ] Editing
           [X] 4-color Printing   [ ] Warehousing/Shipping

Terms: Net 30 on approved credit or by special arrangement.

Mitchell-Shear is a five-year-old division of Lithographics, a general commercial printer. Their prices seem very reasonable, and their work that we have seen is of high quality. They say of themselves, "We are a very small, high quality shop offering good delivery schedules at very competitive prices." Check them out when you are ready to do your next book.

Modern Graphics
P O Box 288
Randolph MA 02368

617-986-4262
John F. Kelly
Vice President

They did not answer our recent printer survey, so we cannot give you any details about their capabilities or services other than that they are supposed to be complete book manufacturers.

Moran Colorgraphic
5425 Florida Blvd / P O 66538
Baton Rouge LA 70896

504-923-2550
Ernest Seals
Vice President

They can do typesetting, printing, and binding. They are also brokers for other book manufacturing services. However, they have not answered our last two RFQ's or printer surveys.

Morgan Press
145 Palisade Street
Dobbs Ferry NY 10522

914-693-0023
Lloyd Morgan
President

Associated with Morgan and Morgan Publishers, they are general commercial printers who will also print limited edition books. They did not answer our recent printer survey, so we cannot give you any other details regarding their capabilities and services.

=====

Morgan Printing
900 Old Koenig Lane #137
Austin TX 78756

512-459-5194
Mike Morgan
President

SP

Quantities: Min: 50   Max: 2500   Opt: 500 - 1000

Book Sizes: 5 1/2 x 8 1/2; 6 x 9; 8 1/2 x 11

Bindings: [I] PB   [I] SS   [O] HC   [I] C/SB

Capabilities: [X] Magazines   [ ] Galley Copies
               [X] Journals   [ ] Demand Printing
               [X] Cookbooks   [ ] 4-color Juvenile Books
               [ ] Yearbooks   [ ] Annual Reports/Brochures
               [X] Catalogs   [ ] Other Commercial Printing

Services: [X] Typesetting   [ ] Teletypesetting
           [ ] Design and Pasteup   [ ] Editing
           [X] 4-color Printing   [ ] Warehousing/Shipping

Terms: 50% down, 50% C.O.D.; 5% discount for prepayment of full amount.

For self-publishers and smaller publishers, Morgan offers very competitive prices on very short runs because they print from paper plates except where 133 half-tones are required. Their work that we have seen is of good quality and appearance. Send for their free price list (they also offer free layout sheets to their customers).

```
* * * * * * * * * * * * * * * * * * * * * * * * * * * * * * * *
*  Tip:  Look around for inexpensive photo and art sources.    *
*        When practical, use stock photo and clip art services *
*        rather than hiring freelancers.  Or use art students  *
*        from your local college.  You can also get excellent  *
*        free photos from your local historical society and    *
*        many corporate or government public relations offices. *
* * * * * * * * * * * * * * * * * * * * * * * * * * * * * * * *
```

Morningrise Printing
1525 W MacArthur Blvd #1
Costa Mesa CA 92626

714-957-8494
Jane McLaughlin
Trustee

Quantities: Min: 250   Max: 5000   Opt: 2000 - 4000

Book Sizes: 5 1/2 x 8 1/2; 8 1/2 x 11

Bindings: [O] PB   [I] SS   [O] HC   [O] C/SB

Capabilities: [ ] Magazines   [ ] Galley Copies
               [ ] Journals   [ ] Demand Printing
               [X] Cookbooks   [X] 4-color Juvenile Books
               [ ] Yearbooks   [ ] Annual Reports/Brochures
               [X] Catalogs   [X] Other Commercial Printing

Services: [X] Typesetting   [ ] Teletypesetting
           [X] Design and Pasteup [ ] Editing
           [ ] 4-color Printing   [ ] Warehousing/Shipping

Terms: 50% down, balance on delivery.

Morningrise is a local printing shop who would like to do more
short-run book printing. They have already done a number of
titles for various customers. You might want to check them out
if you are located near them. They say that they are "concerned
about quality and are very service oriented."

=====

Multiprint Inc.
P O Box 845
Rutherford NJ 07070

201-935-7474
Martin B. Schwalbaum
General Manager

Multiprint used to be a short-run book printer offering very
good prices for small runs. Apparently their prices were too
good since their printing business is now defunct. They are now
printing brokers, calling themselves "The Matchmaker." They
offer a free production guide which is well worth reading.

=====

Murray Printing Company
Pleasant Street
Westford MA 01886

617-692-6321
James F. Conway III
Vice President of Marketing

A subsidiary of Courier Corporation, they have sales offices
in a number of cities (see next page):

130

```
404-451-7262  Atlanta GA          619-271-9224  San Diego CA
312-490-0925  Chicago IL          415-947-0570  San Francisco CA
212-490-8700  New York NY
```

Quantities: Min: 1000   Max: 250,000   Opt: 7000 - 15,000

Book Sizes: almost any size

Bindings: [I] PB   [I] SS   [I] HC   [O] C/SB

Capabilities: [ ] Magazines   [ ] Galley Copies
               [ ] Journals   [ ] Demand Printing
               [ ] Cookbooks   [ ] 4-color Juvenile Books
               [ ] Yearbooks   [ ] Annual Reports/Brochures
               [X] Catalogs   [ ] Other Commercial Printing

Services: [ ] Typesetting   [ ] Teletypesetting
           [ ] Design and Pasteup   [ ] Editing
           [ ] 4-color Printing   [X] Warehousing/Shipping

Terms: Will discuss.

In business since 1897, they are book printing specialists (in one or two colors, casebound or paperback, web or sheet fed). They also do reprints, catalogs, manuals, and workbooks. They emphasize the quality of their work: "We know how to do it right."

User Comment: "Good quality but poor customer service."

RATINGS	1	2	3	4	5	6	7	8	9	10	Ave	
Speed	-	-	-	1	-	-	-	-	-	-	---	1
Price	-	-	-	-	-	1	-	-	-	-	---	
Dependability	-	-	-	1	-	-	-	-	-	-	---	
Service . . .	-	-	-	1	-	-	-	-	-	-	---	
Quality . . .	-	-	-	-	-	-	-	1	-	-	---	
Overall . . .	-	-	-	-	-	1	-	-	-	-	---	

National Publishing Company   215-732-1863
24th & Locust / P O Box 8386   Alan Maxwell
Philadelphia PA 19101

Another subsidiary of the Courier Corporation, they specialize in printing books with lightweight papers and special binding requirements. They also offer editorial services. However, they have never answered any of our RFQ's or printer surveys. They probably do not do short runs.

National Reproductions Corp.   313-961-5252
433 E Larned   Thomas Burkhardt
Detroit MI 48226   Sales Manager

Quantities: Min: 100   Max: 7500   Opt: 1000

Book Sizes: 5 1/2 x 8 1/2; 6 x 9; 8 1/2 x 11

Bindings: [I] PB   [I] SS   [ ] HC   [I] C/SB

Capabilities: [ ] Magazines   [X] Galley Copies
                [X] Journals   [X] Demand Printing
                [X] Cookbooks   [ ] 4-color Juvenile Books
                [ ] Yearbooks   [ ] Annual Reports/Brochures
                [X] Catalogs   [X] Other Commercial Printing

Services: [ ] Typesetting   [ ] Teletypesetting
           [ ] Design and Pasteup [ ] Editing
           [ ] 4-color Printing   [ ] Warehousing/Shipping

Terms: Net 30 with approved credit. Also accepts credit cards.

In business since 1953, National Reproductions specializes in
short-run printing of softcover books. They will also do demand
printing for you (printing only the number of copies you have
orders for and shipping the books direct to your customers).

They actually printed the first edition of this Directory and
did a very fine job -- and delivered very quickly. They are easy
to work with, follow instructions well, and meet their delivery
dates. They combine superb quality with fast, reliable delivery.

User Comment: "Good quality, great prices, and good customer
service. (They quoted me on a saddlestitched book, decided it
wouldn't work, put a perfect binding on, and didn't charge me
extra!) Good company."

RATINGS	1	2	3	4	5	6	7	8	9	10	Ave	
Speed	-	-	-	-	-	-	-	1	1	-	---	2
Price	-	-	-	-	-	-	-	-	1	1	---	
Dependability	-	-	-	-	-	-	-	1	1	-	---	
Service	-	-	-	-	-	-	-	-	2	-	---	
Quality	-	-	-	-	-	-	1	1	-	-	---	
Overall	-	-	-	-	-	-	-	1	1	-	---	

* *
* **Tip:** For some of your publications you may be able to use   *
*   a self-cover (for booklets, manuals, and catalogs).   *
*   You'll save the cost of a separate press run.   *
* *

Naturegraph Publishers Inc.　916-493-5353
3543 Indian Creek Rd / PO 1075　Barbara Brown
Happy Camp　CA 96039　Sales Manager

These people are publishers as well as printers. However, they have never answered any of our printer surveys so we cannot tell you much more about them.

==

Neibauer Press　215-322-6200
20 Industrial Drive　Nathan Neibauer
Warminster PA 18974　President

Quantities: Min: 500　Max: 15,000　Opt: not stated

Book Sizes: 5 1/2 x 8 1/2; 6 x 9; 7 x 10; 8 1/2 x 11

Bindings: [I] PB　[I] SS　[O] HC　[ ] C/SB

Capabilities:　[X] Magazines　　[ ] Galley Copies
　　　　　　[X] Journals　　　[ ] Demand Printing
　　　　　　[X] Cookbooks　　[ ] 4-color Juvenile Books
　　　　　　[ ] Yearbooks　　[X] Annual Reports/Brochures
　　　　　　[X] Catalogs　　　[X] Other Commercial Printing

Services:　[X] Typesetting　　　[X] Teletypesetting
　　　　　[X] Design and Pasteup　[X] Editing
　　　　　[X] 4-color Printing　　[X] Warehousing/Shipping

Terms: Net 30 days after establishing credit.

Neibauer is another general commercial printer who can also print and bind books.

==

Nimrod Press　617-437-7900
170 Brookline Avenue　Walter T. Tower
Boston MA 02215　President

Quantities: Min: 500　Max: 50,000　Opt: 3000

Book Sizes: 5 1/2 x 8 1/2; 6 x 9; 6 1/2 x 9 3/4; 8 1/2 x 11

Bindings: [I] PB　[I] SS　[X] HC　[I] C/SB

For more details, turn the page. Thanks.

Nimrod Press continued

Capabilities: [X] Magazines   [ ] Galley Copies
              [X] Journals   [ ] Demand Printing
              [X] Cookbooks   [X] 4-color Juvenile Books
              [ ] Yearbooks   [X] Annual Reports/Brochures
              [X] Catalogs   [X] Other Commercial Printing

Services: [X] Typesetting   [X] Teletypesetting
            [X] Design and Pasteup   [ ] Editing
            [X] 4-color Printing   [ ] Warehousing/Shipping

Terms: Net 30.

Their typesetting department carries a full line of non-Roman alphabets.

Nobel Book Press   212-777-1300
900 Broadway   Sol Weinreich
New York NY 10003   President

Their optimum print run is 5000 copies. Since, however, they have never answered any of our RFQ's or printer surveys, we cannot give any further details about their capabilities.

Northeast Web Printing   516-454-1600
211 Central Avenue   Chuck Leipham
Farmingdale NY 11735   General Manager

They asked to be listed in the Directory but were too late to get a full listing. Since they appear to be web printers, they are probably more price competitive on longer runs.

Oaks Printing Company   215-759-8511
195 Nazareth Pike   Kenneth H. Oaks
Bethlehem PA 18017   President

Quantities: Min: 10   Max: open   Opt: 5000 - 10,000

Book Sizes: 5 1/2 x 8 1/2; 6 x 9; 8 1/2 x 11

Bindings: [O] PB   [I] SS   [O] HC   [I] C/SB

Capabilities: [X] Magazines   [X] Galley Copies
                [X] Journals   [X] Demand Printing
                [X] Cookbooks   [X] 4-color Juvenile Books
                [ ] Yearbooks   [X] Annual Reports/Brochures
                [X] Catalogs   [X] Other Commercial Printing

Services: [X] Typesetting   [ ] Teletypesetting
            [X] Design and Pasteup   [ ] Editing
            [X] 4-color Printing   [X] Warehousing/Shipping

Terms: Net 30 days.

Offset Composition Services   202-783-1010
419 Seventh St NW #505   F. Scott Watkins
Washington DC 20034   President

They edit manuscripts, design text, and typeset books as well
as print and bind. However we do not know whether or not they do
short runs since they did not answer our RFQ or printer survey.

Offset Paperback Manufacturers   717-675-5261 / 212-489-1557
P O Box N, Route 309N   Robert O'Connor
Dallas PA 18612   Vice President Sales

Offset (which is owned by Bertelsmann, the same German
publishing group that owns Delta Lithograph and Bantam Books) has
been advertising alot recently in Small Press and Writer's Digest
as well as Publishers Weekly. Judging by where they are advertising, you'd think they'd be interested in doing short runs.
But apparently that's not the case. We've sent several queries
on our letterhead as well as twice sending in their ad coupon
requesting more information, but have never heard a peep from
them. Meanwhile, they have not answered our RFQ's or printer
surveys. In their ads they say that "service is priority 1," but
it seems they still have some work to do on their mail handling.

They are one of the hundred largest printers in the United
States, with annual sales over $35,000,000. They are paperback
specialists (trade or mass-market), and can print any quantity
from 5000 to 5,000,000. The quality of their work is very good.
Perhaps by the time the next edition of this Directory comes out,
they will have resolved their mail handling problems.

Omega Industry
Route 2 / Box 28
Reidsville NC 27320

?   SP
Bill Davis Sr
President

They specialize in printing 24 to 48-page poetry chapbooks in
quantities from 10 to 100.

Omnipress
454 W Johnson St / PO Box 7125
Madison WI 53707

800-828-0305   /   608-257-7275
John Vogel
Sales Manager

Quantities: Min: 50   Max: 1000   Opt: 500

Book Sizes: 5 3/8 x 8 1/4; 6 x 9; 8 3/8 x 10 3/4

Bindings: [I] PB   [ ] SS   [ ] HC   [I] C/SB

Capabilities: [ ] Magazines   [ ] Galley Copies
                [X] Journals   [ ] Demand Printing
                [ ] Cookbooks   [ ] 4-color Juvenile Books
                [ ] Yearbooks   [ ] Annual Reports/Brochures
                [ ] Catalogs   [ ] Other Commercial Printing

Services: [ ] Typesetting   [ ] Teletypesetting
            [ ] Design and Pasteup   [ ] Editing
            [ ] 4-color Printing   [ ] Warehousing/Shipping

Terms: Approved accounts available to qualified customers.

They specialize in limited press runs (under 1000) of books,
manuals, directories, and conference proceedings with a high page
count (over 100 pages). They have a completely automated printing and collating system using paper plates. They offer very
quick turnaround (2 - 3 weeks). They can provide 4-color covers.

O'Neil Data Systems
2034 Armacost Avenue
Los Angeles CA 90025

213-820-4247
Marvin Smith
Sales Manager

Quantities: Min: 100   Max: 100,000   Opt: 25,000

Book Sizes: 5 1/2 x 8 1/2; 8 1/2 x 11

Bindings: [I] PB   [I] SS   [ ] HC   [ ] C/SB

Capabilities: [ ] Magazines   [ ] Galley Copies
               [ ] Journals   [ ] Demand Printing
               [ ] Cookbooks   [ ] 4-color Juvenile Books
               [ ] Yearbooks   [ ] Annual Reports/Brochures
               [X] Catalogs   [X] Other Commercial Printing

Services: [X] Typesetting   [ ] Teletypesetting
           [ ] Design and Pasteup [ ] Editing
           [ ] 4-color Printing   [ ] Warehousing/Shipping

Terms: Net 30 days.

   They specialize in typesetting and printing price catalogs,
inventory lists, parts manuals, reference materials, membership
directories, and other publications whose data is stored on a
computer. They can produce press-ready plates direct from
magnetic tapes or floppy discs; hence, they can save you time and
money on pre-press services. They can print and deliver in four
days.

Optic Graphics Inc.   800-638-7107 / 301-768-3000
101 Dover Road   John Hamlett
Glen Burnie MD 21061   Senior Estimator

   In Washington, DC call 202-261-1540.

Quantities: Min: 500   Max: open   Opt: 5000 - 10,000

Book Sizes: almost any trim size

Bindings: [I] PB   [ ] SS   [I] HC   [I] C/SB

Capabilities: [ ] Magazines   [ ] Galley Copies
               [ ] Journals   [ ] Demand Printing
               [ ] Cookbooks   [ ] 4-color Juvenile Books
               [ ] Yearbooks   [X] Annual Reports/Brochures
               [ ] Catalogs   [ ] Other Commercial Printing

Services: [ ] Typesetting   [ ] Teletypesetting
           [ ] Design and Pasteup [ ] Editing
           [ ] 4-color Printing   [ ] Warehousing/Shipping

Terms: 2% 10, net 30 with approved credit.

   They can manufacture both vinyl and cloth looseleaf binders
in-house. They assemble binders and/or slipcases as well as
print manuals for the computer software industry. According to
one of their customers, their prices are "great."

137

Oxford Group
P O Box 269
Norway ME 04268

207-743-8953
Howard James
President

They have another printing plant at 351 Main Street, P. O. Box 38, Berlin, NH 03570; phone 603-752-2339.

According to their listing in the LMP they do short runs, but they haven't answered our last two RFQ's or printer surveys. Maybe they are only interested in working with local customers. Check them out if you are located in New England.

Oxmoor Press
100 2 Oxmoor Rd / P O Box 980
Birmingham AL 35201

205-942-0511
James Allen
Sales Manager

Oxmoor, a subsidiary of Stevens Graphics and associated with Oxmoor Publishing Company, prints lightweight papers on either offset or letterpress. They can print Bibles, catalogs, directories, even coloring books. We're not sure, however, whether or not they do short runs. They have not answered our printer surveys.

Pantagraph Printing
217 W Jefferson St / P O 1406
Bloomington IL 61701

309-829-1071
Fred Dolan
President

They have never answered any of our RFQ's or survey forms. They apparently do not do short runs.

```
* * * * * * * * * * * * * * * * * * * * * * * * * * * * * * * *
*  Tip:  You can save money on your printing bills by editing  *
*        your books more thoroughly.  Does the book really     *
*        have to be as long as it is?  Can some chapters be     *
*        trimmed or even deleted without hurting the content,  *
*        design, and message of the book?  Some books would be *
*        better off with some judicious editing.  Not only     *
*        would the content be improved, but the book would     *
*        actually sell better.  Is your book overly verbose,   *
*        unwieldy, pretentious?  Be honest now.  Do yourself,  *
*        your authors, and your customers a favor by aggres-   *
*        sively editing your books -- to make them better.     *
* * * * * * * * * * * * * * * * * * * * * * * * * * * * * * * *
```

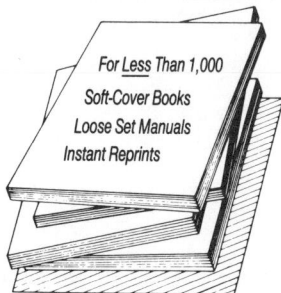

Paraclete Press
5 Bay View Dr / P O Box 1568
Orleans MA 02653

617-255-4685
Robert Jamison
Sales Manager

Quantities: Min: 500   Max: 25,000   Opt: 3000

Book Sizes: 5 1/2 x 8 1/2; 6 x 9; 8 1/2 x 11

Bindings: [I] PB   [I] SS   [ ] HC   [ ] C/SB

Capabilities: [X] Magazines   [X] Galley Copies
               [ ] Journals   [ ] Demand Printing
               [ ] Cookbooks   [X] 4-color Juvenile Books
               [ ] Yearbooks   [X] Annual Reports/Brochures
               [X] Catalogs   [X] Other Commercial Printing

Services: [X] Typesetting   [ ] Teletypesetting
           [X] Design and Pasteup   [ ] Editing
           [X] 4-color Printing   [ ] Warehousing/Shipping

Terms: Net 30.

A publisher who also prints for other small presses, Paraclete features total in-house capability for 4-color work -- color separations, typesetting, pasteup, printing, and binding.

===

Parthenon Press
201 Eighth Avenue S
Nashville TN 37202

800-251-4857 / 615-749-6464
Nathan Honeycutt
Sales Manager

Associated with Abingdon Press and the United Methodist Publishing House, Parthenon's minimum press run is 20,000 copies.

===

Patterson Printing Company
1550 Territorial Rd / P O 1244
Benton Harbor MI 49022

616-925-2177
Leroy Patterson
President

Quantities: Min: 200   Max: 100,000+   Opt: 5000 - 10,000

Book Sizes: 5 1/2 x 8 1/2; 6 x 9; 8 1/2 x 11

Bindings: [I] PB   [I] SS   [O] HC   [I] C/SB

For more details, see the page following their ad.

PATTERSON PEOPLE DELIVER!

Obviously we have all the latest equipment and technology to produce our special products. When you serve distinguished publishers who demand true quality in their educational texts, course books, student workbooks, spirit masters, and a wide variety of directories and technical manuals, you must have the latest, most efficient sheet fed, web and bindery equipment available.

But it's our people, not the machines, that actually deliver the goods. It's our sales consultants, estimators, and production people who handle those small budgets, short schedules and direction changes, day in and day out, and deliver a quality product every time – on time!

How do they do it? They're experienced professionals… **and they CARE!**

If your printing needs require people with expertise **who CARE,** phone us at **(616) 925-2177.**

PATTERSON PRINTING

Production Printers to the publishers of informational materials

1550 TERRITORIAL ROAD • BENTON HARBOR, MICHIGAN • 49022 • (616) 925-2177

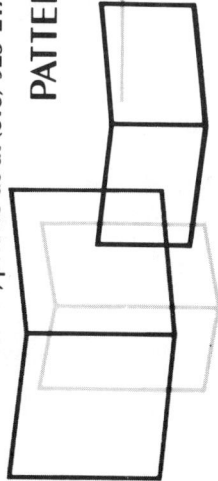

Capabilities: [X] Magazines   [ ] Galley Copies
              [X] Journals   [ ] Demand Printing
              [ ] Cookbooks   [ ] 4-color Juvenile Books
              [ ] Yearbooks   [ ] Annual Reports/Brochures
              [ ] Catalogs   [X] Other Commercial Printing

Services: [ ] Typesetting   [ ] Teletypesetting
           [ ] Design and Pasteup [ ] Editing
           [X] 4-color Printing   [X] Warehousing/Shipping

Terms: Net 30 with approved credit.

Patterson offers very competitive prices for quantities between 5000 and 10,000. They quoted the lowest prices we received
for an 80-page, 8 1/2 x 11 perfectbound paperback.

Their specialty is serving publishers of information materials
(educational textbooks, course books, student workbooks, and
technical manuals). They can also provide fulfillment services.

User Comments: "My only complaint is that the darkness of the
printed pages varies somewhat." ... "They gave me the best bid
out of about 15 printers, were fast, helpful, and high quality."

RATINGS	1	2	3	4	5	6	7	8	9	10	Ave	
Speed	-	-	-	-	-	1	-	-	-	1	---	2
Price	-	-	-	-	-	-	-	-	1	1	---	
Dependability	-	-	-	-	-	1	-	-	1	-	---	
Service . . .	-	-	-	-	-	1	-	-	1	-	---	
Quality . . .	-	-	-	-	1	-	-	-	1	-	---	
Overall . . .	-	-	-	-	-	1	-	-	1	-	---	

==

Paust Incorporated   317-962-1507
14 N 10th St / P O Box 1326   Roland Paust
Richmond IN 47375   Sales Manager

Quantities: Min: 10   Max: open   Opt: 1000 - 5000

Book Sizes: 4 1/4 x 7; 5 1/2 x 8 1/2; 6 x 9; 8 1/2 x 11

Bindings: [O] PB   [I] SS   [O] HC   [I] C/SB

Capabilities: [X] Magazines   [X] Galley Copies
              [X] Journals   [X] Demand Printing
              [ ] Cookbooks   [ ] 4-color Juvenile Books
              [ ] Yearbooks   [X] Annual Reports/Brochures
              [X] Catalogs   [X] Other Commercial Printing

Services: [X] Typesetting   [ ] Teletypesetting
           [X] Design and Pasteup [ ] Editing
           [X] 4-color Printing   [ ] Warehousing/Shipping

Terms: Typesetting: 1/2 with order, balance with returned proofs.
        Printing: payment in full with order.

In business since 1945, they offer personalized graphic arts
services. They can print direct mail materials, industrial catalogs, advertising brochures, point of sale displays, and annual
reports as well as books.

Pearl Pressman Liberty Printing   215-925-4900 / 212-925-5162
Fifth and Poplar Streets   Marvin Bergman
Philadelphia PA 19123   Sales Manager, Book Division

They specialize in printing multicolor medical books and juvenile picture books. They did not answer our RFQ or printer
survey, so we cannot provide any other details about their capabilities or services.

Pelican Pond Publishing   ?
13386 N Bloomfield Road   Attn: President
Nevada City CA 95959

Pelican was recommended to us by one of their customers; however they did not answer the printer survey we sent them, so we
cannot tell you any further details. They are apparently distri-
butors and publishers as well.

Pennysaver Press Inc.   516-997-7755
50 Jericho Turnpike   Steve Ferber
Jericho NY 11753   General Manager

Quantities: Min: 5000   Max: 1,000,000 Opt: 10,000 - 50,000

Book Sizes: 7 x 10; 7 1/4 x 10 3/4; 8 3/8 x 10 3/4; 8 1/2 x 11

For more information about their capabilities and services,
turn the page. Thanks again.

Pennysaver Press Inc. continued

Bindings: [O] PB   [I] SS   [ ] HC   [ ] C/SB

Capabilities: [X] Magazines   [ ] Galley Copies
               [ ] Journals   [ ] Demand Printing
               [ ] Cookbooks   [ ] 4-color Juvenile Books
               [ ] Yearbooks   [ ] Annual Reports/Brochures
               [X] Catalogs   [X] Other Commercial Printing

Services: [X] Typesetting   [ ] Teletypesetting
           [X] Design and Pasteup [ ] Editing
           [ ] 4-color Printing   [X] Warehousing/Shipping

Terms: Negotiable.

They specialize in low-cost web offset printing of catalogs,
newspapers, circulars, booklets, newsletters, comic books, and
books. They offer their best prices in quantities over 10,000.

===

Phillips Brothers Printing   800-637-9444 / 217-787-3014
1555 W Jefferson / P O Box 580   Roger Walker
Springfield IL 62705   Sales Representative

Quantities: Min: 1000 Max: 500,000   Opt: 10,000

Book Sizes: 4 1/4 x 8; 5 1/2 x 8 1/2; 6 x 9; 8 x 9 1/4;
             8 1/2 x 11

Bindings: [I] PB   [I] SS   [O] HC   [I] C/SB

Capabilities: [ ] Magazines   [ ] Galley Copies
               [X] Journals   [ ] Demand Printing
               [X] Cookbooks   [X] 4-color Juvenile Books
               [ ] Yearbooks   [X] Annual Reports/Brochures
               [X] Catalogs   [X] Other Commercial Printing

Services: [ ] Typesetting   [ ] Teletypesetting
           [ ] Design and Pasteup [ ] Editing
           [X] 4-color Printing   [ ] Warehousing/Shipping

Terms: Net 30 days.

A family-owned business since 1883, Phillips specializes in
web offset printing of books and catalogs. Just because they're
old, however, doesn't mean they're outdated. They've just
recently moved into new modernized facilities. They, too, offer
their best prices and services on print runs over 10,000 copies.

Plain Talk Publishing Company   515-282-0483
511 E Sixth Avenue   John D. Mertz
Des Moines   IA 50309   Vice President Sales

Quantities: Min: 10   Max: 25,000   Opt: 10,000

Book Sizes: almost any size

Bindings: [O] PB   [I] SS   [O] HC   [I] C/SB

Capabilities: [X] Magazines   [X] Galley Copies
               [X] Journals   [X] Demand Printing
               [ ] Cookbooks   [ ] 4-color Juvenile Books
               [ ] Yearbooks   [X] Annual Reports/Brochures
               [X] Catalogs   [X] Other Commercial Printing

Services: [X] Typesetting   [ ] Teletypesetting
           [X] Design and Pasteup   [ ] Editing
           [X] 4-color Printing   [X] Warehousing/Shipping

Terms: To be negotiated.

Plain Talk specializes in 4-color catalogs, brochures, annual reports, and general commercial printing.

Plus Communications   ?
1044 Pershall Road   Attn: President
St Louis   MO 63137

Plus was recommended to us by one of their customers; however, they did not answer our printer survey so we cannot give you any other details concerning their services and capabilities.

Port City Press   301-486-3000
1323 Greenwood Road   Robert H. Cooper
Baltimore   MD 21208   Vice President

They have sales offices in the following places:

617-482-4226 Boston MA   212-921-9166 New York NY
609-858-3988 Delaware Valley   202-635-1200 Washington DC

For complete details on their capabilities and services, turn the page.

145

Port City Press continued

Quantities: Min: 500   Max: 100,000   Opt: 10,000 - 50,000

Book Sizes: 5 1/2 x 8 1/2; 6 x 9; 7 x 10; 8 1/2 x 11

Bindings: [I] PB   [I] SS   [ ] HC   [ ] C/SB

Capabilities: [ ] Magazines   [ ] Galley Copies
                [ ] Journals   [ ] Demand Printing
                [ ] Cookbooks   [ ] 4-color Juvenile Books
                [ ] Yearbooks   [ ] Annual Reports/Brochures
                [ ] Catalogs   [ ] Other Commercial Printing

Services: [X] Typesetting   [ ] Teletypesetting
            [X] Design and Pasteup   [ ] Editing
            [X] 4-color Printing   [X] Warehousing/Shipping

Terms: Net upon receipt of invoice with approved credit.

In business since 1961, Port City specializes in printing and binding softcover books. They offer their best prices on runs between 10,000 and 50,000.

RATINGS	1	2	3	4	5	6	7	8	9	10	Ave	
Speed	-	-	-	-	-	-	1	-	-	-	---	1
Price	-	-	-	-	-	1	-	-	-	-	---	
Dependability	-	-	-	-	-	-	-	-	1	-	---	
Service . . .	-	-	-	-	-	-	-	-	1	-	---	
Quality . . .	-	-	-	-	-	-	-	-	1	-	---	
Overall . . .	-	-	-	-	-	-	-	-	1	-	---	

PPI Press   212-292-5536 / 914-738-3810
940 E 149th Street   Sam Silberberg
Bronx NY 10455   President

They did not answer our recent printer survey.

Premier Printing Corporation   714-871-3121 / 213-691-4133
124 W Wilshire Ave / Box 348   Dana Cordrey
Fullerton CA 92632   President

Quantities: Min: 500   Max: 20,000   Opt: 5000 - 10,000

Book Sizes: 5 1/2 x 8 1/2; 6 x 9; 8 1/2 x 11

Bindings: [O] PB   [I] SS   [O] HC   [O] C/SB

Capabilities: [X] Magazines   [ ] Galley Copies
               [X] Journals   [ ] Demand Printing
               [ ] Cookbooks   [ ] 4-color Juvenile Books
               [ ] Yearbooks   [X] Annual Reports/Brochures
               [X] Catalogs   [X] Other Commercial Printing

Services: [X] Typesetting   [X] Teletypesetting
           [X] Design and Pasteup   [ ] Editing
           [X] 4-color Printing   [ ] Warehousing/Shipping

Terms: 1 1/2% discount 10 days, net 30 days with approved credit.

   Also known as Sultana Press, Premier can only provide saddle-
stitch binding in-house; all other bindings are subcontracted.

═══

Prinit Press   317-478-4885   SP
2151 Franklin St / P O Box 65   Bob Goodwin
Dublin IN 47335   President

Quantities: Min: 250   Max: 10,000   Opt: 2000 - 5000

Book Sizes: 4 1/4 x 7; 5 1/2 x 8 1/2; 8 1/2 x 11

Bindings: [I] PB   [I] SS   [O] HC   [O] C/SB

Capabilities: [X] Magazines   [ ] Galley Copies
               [ ] Journals   [ ] Demand Printing
               [X] Cookbooks   [ ] 4-color Juvenile Books
               [ ] Yearbooks   [ ] Annual Reports/Brochures
               [X] Catalogs   [X] Other Commercial Printing

Services: [X] Typesetting   [ ] Teletypesetting
           [X] Design and Pasteup   [ ] Editing
           [X] 4-color Printing   [ ] Warehousing/Shipping

Terms: 1/3 down, 1/3 in 45 days, 1/3 due on completion.
        Established accounts: 2% 15, net 30.

   Prinit has been advertising in Writer's Digest for years and
is accustomed to working with self-publishers. Send for their
free price list. Their prices are better than those offered by
Adams Press, which caters to the same clientele. Prinit guaran-
tees (in writing) the quality of their workmanship -- try getting
that from any other printer.

The Print Center
P O Box 1050
Brooklyn NY 11202

212-206-8465
Robert Hershon
Executive Director

Quantities: Min: 250   Max: 10,000   Opt: 1000 - 5000

Book Sizes: almost any size

Bindings: [X] PB   [X] SS   [X] HC   [X] C/SB

Capabilities: [X] Magazines   [X] Galley Copies
                 [X] Journals   [ ] Demand Printing
                 [ ] Cookbooks   [ ] 4-color Juvenile Books
                 [ ] Yearbooks   [X] Annual Reports/Brochures
                 [ ] Catalogs   [ ] Other Commercial Printing

Services: [X] Typesetting   [ ] Teletypesetting
             [X] Design and Pasteup [ ] Editing
             [X] 4-color Printing   [ ] Warehousing/Shipping

Terms: Varies with situation, project, etc.

The Print Center is a non-profit corporation serving non-commercial publishers of literature, the arts, and work of public interest. They also work with individual writers and artists.

Printing Corporation of America
620 SW 12th Avenue
Pompano Beach FL 33060

305-781-8100
Murray Tuchman
President

Quantities: Min: 100   Max: 40,000   Opt: 5000 - 25,000

Book Sizes: 4 x 6; 5 1/2 x 8 1/2; 6 x 9; 8 1/2 x 11; 9 x 12

Bindings: [I] PB   [I] SS   [O] HC   [I] C/SB

Capabilities: [X] Magazines   [ ] Galley Copies
                 [X] Journals   [X] Demand Printing
                 [X] Cookbooks   [X] 4-color Juvenile Books
                 [ ] Yearbooks   [X] Annual Reports/Brochures
                 [X] Catalogs   [X] Other Commercial Printing

Services: [X] Typesetting   [ ] Teletypesetting
             [X] Design and Pasteup [ ] Editing
             [X] 4-color Printing   [X] Warehousing/Shipping

Terms: First order: 50% down, balance C.O.D.; net 30 on
        subsequent orders subject to credit approval.

Progressive Typographers
P O Box 278
Emigsville PA 17318

717-764-5908
David Rae
General Manager, Printing

They can print and bind saddlestitched publications. Of
course, they also offer complete typesetting services. No other
details of their operation are available since they did not
answer our recent printer survey.

Publishers Choice Book Mfg.
Mars Industrial Pk / P O 848
Mars PA 16046

412-625-3555 / 412-784-1371
Michael D. Cheteyan II
President

Quantities: Min: 500   Max: open   Opt: 65,000

Book Sizes: 5 1/2 x 8 1/2; 6 x 9; 8 1/2 x 11

Bindings: [I] PB   [I] SS   [O] HC   [O] C/SB

Capabilities: [X] Magazines   [ ] Galley Copies
                 [X] Journals   [ ] Demand Printing
                 [ ] Cookbooks   [ ] 4-color Juvenile Books
                 [ ] Yearbooks   [X] Annual Reports/Brochures
                 [X] Catalogs   [X] Other Commercial Printing

Services: [X] Typesetting   [ ] Teletypesetting
            [X] Design and Pasteup   [ ] Editing
            [X] 4-color Printing   [X] Warehousing/Shipping

Terms: Not stated.

They specialize in printing conference proceedings, journals,
publication catalogs, membership directories, and other bound
publications for associations. They also offer complete article
reprint services.

```
* * * * * * * * * * * * * * * * * * * * * * * * * * * * * * * *
*  Tip:  Again, we cannot emphasize enough, it is important to  *
*        set a firm publishing schedule and, in turn, a firm    *
*        production schedule -- with plenty of leeway to allow   *
*        you time to make changes if they are needed.  Set a    *
*        realistic schedule, put it in writing, and stick to    *
*        it.  Don't rush yourself.  Rush jobs are sloppy jobs.  *
*        And sloppy jobs cost you money.  Either you must pay    *
*        to correct the mistakes, or if you don't correct them, *
*        you can lose sales because the book is not suitable.   *
* * * * * * * * * * * * * * * * * * * * * * * * * * * * * * * *
```

Publishers Press
1900 West 2300 South, Box 27408
Salt Lake City UT 84119-0408

801-972-6600
Brad Airmet
Sales Representative

Quantities: Min: 1000   Max: 50,000   Opt: 5000

Book Sizes: any size from 4 x 7 to 9 x 12

Bindings: [I] PB   [I] SS   [I] HC   [I] C/SB

Capabilities: [X] Magazines   [ ] Galley Copies
                [X] Journals   [ ] Demand Printing
                [X] Cookbooks   [ ] 4-color Juvenile Books
                [ ] Yearbooks   [X] Annual Reports/Brochures
                [ ] Catalogs   [X] Other Commercial Printing

Services: [ ] Typesetting   [ ] Teletypesetting
           [ ] Design and Pasteup   [ ] Editing
           [X] 4-color Printing   [X] Warehousing/Shipping

Terms: Net 30 with approved credit; otherwise, 50% down, balance
        on delivery.

Publishers Press can print any size book from very small to
very large with any sort of binding and in one, two, or four
colors. They produce books of superb quality. Indeed, they say
that "quality and service are still our priorities," and "we
pride ourselves on the number of repeat customers we have." They
seem to be very flexible and willing to work with you to produce
the best book possible.

Publishing Resources Inc.
P O Box 41307
San Juan PR 00940

809-724-0318
Anne Chevako
Marketing Manager

They did not answer our survey forms. We are not sure if they
are short-run book printers or simply printing brokers.

Quality Press
3962 S. Mariposa
Denver CO 80223

303-761-2160
Sales Representative

They have never answered any of our RFQ's or survey forms.
They apparently do not do short runs.

Quinn-Woodbine
P O Box 515
Woodbine NJ 08270

609-861-5352 / 212-889-0552
Bob Abrams
Vice-President

Quantities: Min: 200   Max: 5000   Opt: 1000

Book Sizes: 5 1/2 x 8 1/2; 6 x 9; 7 x 10; 8 1/2 x 11

Bindings: [I] PB   [I] SS   [I] HC   [O] C/SB

Capabilities: [ ] Magazines   [ ] Galley Copies
                [X] Journals   [ ] Demand Printing
                [ ] Cookbooks   [ ] 4-color Juvenile Books
                [ ] Yearbooks   [ ] Annual Reports/Brochures
                [ ] Catalogs   [ ] Other Commercial Printing

Services: [ ] Typesetting   [ ] Teletypesetting
            [ ] Design and Pasteup [ ] Editing
            [ ] 4-color Printing   [ ] Warehousing/Shipping

Terms: Net 30 with approved credit.

   Quinn-Woodbine specializes in printing perfectbound and casebound books in short runs from 75 to 5000 copies. They are "very
competitive on smyth-sewn hardcovers" (indeed, they do hardcover
binding for other printers as well).

===

Quintessence Press
356 Bunker Hill Mine Road
Amador City CA 95601

209-267-5470
Linomarl (Marlan Beilke)

Quantities: Min: 300   Max: 5000   Opt: 1500

Book Sizes: miniatures, plus the three standard sizes

Bindings: [I] PB   [ ] SS   [ ] HC   [ ] C/SB

Capabilities: [ ] Magazines   [X] Galley Copies
                [X] Journals   [X] Demand Printing
                [X] Cookbooks   [X] 4-color Juvenile Books
                [ ] Yearbooks   [X] Annual Reports/Brochures
                [X] Catalogs   [X] Other Commercial Printing

Services: [L] Typesetting   [ ] Teletypesetting
            [X] Design and Pasteup [X] Editing
            [ ] 4-color Printing   [ ] Warehousing/Shipping

   For more details, turn the page. Thanks.

Quintessence Press continued

Terms: By arrangement.

They specialize in letterpress printing. "For those still interested in top-flight work, our unusual all-letterpress plant is the alternative to mediocrity at liveable prices." They are the largest linotype/ludlow hot-metal composition house on the West Coast.

Quixott Press
Church School Rd, RR#4
Doylestown PA 18901

215-794-7107
Charles Ingerman
President

They were listed in the last edition of this <u>Directory</u> but did not answer our current survey. They may no longer be interested in doing short-run book printing.

Rae Publishing Company
282 Grove Avenue
Cedar Grove NJ 07009

201-239-1600
John L. Sullivan
Sales Manager

Apparently just a book binder and not a printer as well, Rae won a number of 1984 AIGA awards for their binding work. They did not answer our printer survey, so we cannot give you any further details.

Rand McNally & Company
Book Mfg. Div. / P O Box 7600
Chicago IL 60680

312-267-6868
Paul Alms
Sales Manager

Although a publisher themselves, Rand McNally also offers printing and binding services to other publishers. Indeed, Rand McNally, with annual sales over $150,000,000, is one of the fifteen largest printers in the United States. They not only print atlases and maps, but also encyclopedias, cookbooks, dictionaries, school textbooks, and direct mail books. Apparently, though, they do not do short runs (they did not answer our printer survey requests). They have been in business for almost 130 years.

Readi Multi-Lith
165 S Union Boulevard
Denver CO 80228

303-987-2338
Larry E. Rush
Sales Manager

East Coast publishers may call their New Jersey sales office:
609-596-3003.

Quantities: Min: 10   Max: 1000   Opt: 300

Book Sizes: 8 1/2 x 11 specialists

Bindings: [I] PB   [ ] SS   [ ] HC   [ ] C/SB

Capabilities: [ ] Magazines   [ ] Galley Copies
               [X] Journals   [X] Demand Printing
               [ ] Cookbooks   [ ] 4-color Juvenile Books
               [ ] Yearbooks   [ ] Annual Reports/Brochures
               [X] Catalogs   [ ] Other Commercial Printing

Services: [X] Typesetting   [X] Teletypesetting
           [ ] Design and Pasteup   [ ] Editing
           [ ] 4-color Printing   [ ] Warehousing/Shipping

Other services: database management and microforms

Terms: Net 30 days.

Readi specializes in short runs of large size (8 1/2 x 11)
high-page-count books, directories, and manuals.

Recorder Sunset Press
99 S Van Ness Ave
San Francisco CA 94103

415-621-5400
Paul McDonald
Vice-President

Quantities: Min: 8000   Max: 1,000,000 Opt: 100,000

Book Sizes: 5 1/2 x 8 1/2; 8 1/2 x 11

Bindings: [I] PB   [I] SS   [I] HC   [I] C/SB

Capabilities: [ ] Magazines   [ ] Galley Copies
               [ ] Journals   [ ] Demand Printing
               [ ] Cookbooks   [ ] 4-color Juvenile Books
               [ ] Yearbooks   [ ] Annual Reports/Brochures
               [ ] Catalogs   [ ] Other Commercial Printing

For more about Recorder's services, please turn to the next
page.

Recorder Sunset Press continued

Services: [X] Typesetting   [X] Teletypesetting
           [ ] Design and Pasteup [ ] Editing
           [X] 4-color Printing   [X] Warehousing/Shipping

Terms: Net 30 days with approved credit.

They are "the only printer in San Francisco with a full heat-set web."

Regency Graphics   212-867-5230
501 Fifth Avenue   Attn: Sales Representative
New York NY 10017

They were interested in being listed in this edition of the Directory, but they did not return the survey form we sent them so we can give you no further details regarding their services.

Regensteiner Press   312-666-4200
1224 W Van Buren Street   Frederick Glazer
Chicago IL 60607   President

Apparently they offer pamphlet binding only. They have never answer our RFQ's or printer surveys, so we do not know anything else about them.

Repro-Tech   201-785-0011
250 Lackawanna Ave / P O 600 Howard Herman
West Patterson   NJ 07424

They are printers of business manuals, specializing in runs between 500 and 5000. They've done printing for a number of Fortune 500 companies.

```
* * * * * * * * * * * * * * * * * * * * * * * * * * * * * * *
*  Tip:  Avoid last minute scrambles by leaving enough leeway  *
*        in your production schedule for inevitable delays.     *
* * * * * * * * * * * * * * * * * * * * * * * * * * * * * * *
```

Rich Printing Company
7131 Centennial Blvd, Box 90472
Nashville TN 37209

615-385-3500
James Thomason
President

In business since 1902, they can produce one, two, and four color books, catalogs, and other publications. Since they did not answer our last two RFQ's and printer surveys, no other details are available (they did provide a quote for us on 1000 copies of a 52-page manual, so they can do short runs).

Rollins Press Inc.
1624 Forsyth Road
Orlando FL 32807

305-677-5533
Robert D. Cowart
President

Apparently not a short-run printer, they have never answered our RFQ's or printer surveys.

Ronalds Federated Ltd.
150 Bloor St W #805
Toronto, Ontario
M5S 2X9   Canada

416-964-1374
Frank Rolph
President

Ronalds, one of the largest Canadian printers with annual sales over $80,000,000, is primarily a commercial printer who can also produce softcover books. Again, no other details are available since they did not reply to our printer surveys.

Rose Printing Company
2503 Jackson Bluff / P O 5078
Tallahassee FL 32314

904-576-4151
Richard G. Walsh
Marketing Director

They have sales offices in the following locations:

301-889-6110 Baltimore MD
212-986-7282 New York NY

305-352-0590 Orlando FL
202-889-6110 Washington DC

Quantities: Min: 500   Max: 100,000   Opt: 5000 – 20,000

For more details about their capabilities and services, turn the page.

Rose Printing Company continued

Book Sizes: 3 1/2 x 5 1/2; 5 1/2 x 8 1/2; 6 x 9; 8 1/2 x 11

Bindings: [I] PB   [I] SS   [I] HC   [I] C/SB

Capabilities: [X] Magazines   [ ] Galley Copies
                [X] Journals   [X] Demand Printing
                [X] Cookbooks   [X] 4-color Juvenile Books
                [ ] Yearbooks   [X] Annual Reports/Brochures
                [ ] Catalogs   [ ] Other Commercial Printing

Services: [X] Typesetting   [X] Teletypesetting
            [X] Design and Pasteup   [ ] Editing
            [X] 4-color Printing   [ ] Warehousing/Shipping

Terms: Net 30 with approved credit. Otherwise, 1/3 with order,
        1/3 with press proofs, and 1/3 before shipment.

   In business for over 50 years, Rose has advertised a standard
book format that could save you money if your book meets these
criteria: 5 1/2 x 8 1/2 or 6 x 9; perfectbound; black ink text;
50 lb. or 60 lb. white offset or 60 lb. natural; 10 pt C1S cover
in one to four colors, varnish coating, bulk packing in cartons.
Their other speciality is high quality casebound juvenile books.

   Besides regular trade books, they also print 3 1/2 x 5 1/2
minibooks (purse books), advertising card decks, stamp saver
books, and airline tickets. They can print these in quantities
up to a million or more.

   They can provide you with a computerized quote in fifteen
minutes. Their usual turnaround time on books is four to six
weeks.

   User Comments: "There were white spots on some of the covers
of the first printing. The second printing of my book had all
4,200 copies with their spines wrinkled."... "Did a nice job for
us on an 8 1/2 x 11 looseleaf and hardbound book in 1983 at
competitive prices."

RATINGS	1	2	3	4	5	6	7	8	9	10	Ave	
Speed	-	-	-	-	1	-	-	-	-	-	---	1
Price	-	-	-	-	-	-	1	-	-	-	---	
Dependability	-	-	1	-	-	-	-	-	-	-	---	
Service	-	-	-	1	-	-	-	-	-	-	---	
Quality	1	-	-	-	-	-	-	-	-	-	---	
Overall	-	-	1	-	-	-	-	-	-	-	---	

*** Use this <u>Directory</u> to help you select the most likely
    printers to query for each book you want to publish.

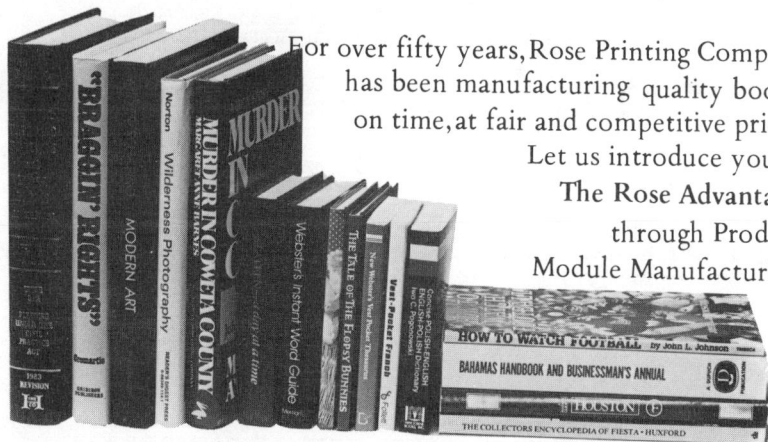

The Saybrook Press
146 Elm Street / P O Box 629
Old Saybrook CT 06475

203-388-5737
William M. Pike
President

They say they specialize in short runs, but they have never
answered any of our RFQ's or printer surveys. Perhaps they only
want business from the New England area.

Schiff Printers & Lithographer
1107 Washington Blvd / P O 5131
Pittsburg PA 15206

412-441-5760
Scott Barthelmes
Customer Service

Quantities: Min: 1000   Max: 50,000   Opt: 5000

Book Sizes: 5 1/2 x 8 1/2; 6 x 9; 8 1/2 x 11; other sizes

Bindings: [O] PB   [I] SS   [O] HC   [I] C/SB

Capabilities: [X] Magazines   [X] Galley Copies
                [X] Journals   [ ] Demand Printing
                [ ] Cookbooks   [ ] 4-color Juvenile Books
                [ ] Yearbooks   [X] Annual Reports/Brochures
                [X] Catalogs   [X] Other Commercial Printing

Services: [X] Typesetting   [X] Teletypesetting
            [X] Design and Pasteup   [ ] Editing
            [X] 4-color Printing   [ ] Warehousing/Shipping

Terms: Not stated.

A full-service commercial printing firm, Schiff offers typesetting, full graphics support (with a creative art director and
keyline artists), letterpress and offset printing, and in-house
saddlestitching and comb-binding. Other book bindings can be
sub-contracted. They are also equipped to produce newsletters,
bulletins, product literature, brochures, and folders in one to
four colors.

G. Schirmer Inc.
48-02 48th Avenue
Woodside NY 11377

212-784-8520
William Bunchuck
Sales Manager

They apparently do not do short-run book printing since they
did not answer our recent printer surveys.

Schlasbach Printers
RR 1 / Box 301
Sugarcreek OH 44681

?
Attn: President

We list these printers because they were rated by one of the publishers responding to our user survey (see below). Schlasbach did not, however, respond to the printer survey we sent them so we cannot give you any other details regarding their services and capabilities.

RATINGS	1	2	3	4	5	6	7	8	9	10	Ave	
Speed	-	-	-	-	-	-	1	-	-	-	---	1
Price	-	-	-	-	-	-	-	-	1	-	---	
Dependability	-	-	-	-	-	-	-	-	-	-	---	
Service	-	-	-	-	-	-	-	-	-	-	---	
Quality	-	-	-	-	-	-	-	-	-	-	---	
Overall	-	-	-	-	-	1	-	-	-	-	---	

Science Press
300 W Chestnut Street
Ephrata PA 17522

717-733-7981
Volker Kruhoeffer
Vice President Sales & Marketing

A division of Monroe Printing Company, Science Press did not answer our survey forms, nor return our five phone calls. Perhaps you will have better luck with their New York or Washington, DC sales offices: New York (212-661-0786) and Washington, DC (703-450-4477). They print professional journals, directories, catalogs, books, and commercial materials in short to medium runs. They do printing for many university and association presses.

Scribner Graphic Press
2580 Wyandotte #B
Mountain View CA 94043

415-967-8118
Judith A. McKim
Account Executive

Quantities: Min: 500   Max: 50,000   Opt: 5000 - 30,000

Book Sizes: 5 1/2 x 8 1/2; 6 x 9; 8 1/2 x 11; and others

Bindings: [O] PB   [I] SS   [ ] HC   [I] C/SB

For more details on Scribner's capabilities and services, please turn the page.

Capabilities: [X] Magazines   [ ] Galley Copies
              [X] Journals   [ ] Demand Printing
              [X] Cookbooks   [ ] 4-color Juvenile Books
              [ ] Yearbooks   [X] Annual Reports/Brochures
              [X] Catalogs   [X] Other Commercial Printing

Services: [X] Typesetting   [ ] Teletypesetting
           [ ] Design and Pasteup   [ ] Editing
           [X] 4-color Printing   [X] Warehousing/Shipping

Terms: Net 30 with approved credit.

Scribner is a commercial printer specializing in short to medium runs of two-color booklets, catalogs, manuals, newsletters and other similar multiple-page publications. They offer "quick turnaround" and "excellent quality."

═══

Semline Inc.   617-848-2380
180 Wood Road   John T. Collins
Braintree MA 02184   President

They have branch offices in Washington, DC; New York, NY; Westwood, MA; and Bridgewater, MA.

Quantities: Min: 1000   Max: open   Opt: 10,000 - 50,000

Book Sizes: 5 1/2 x 8 1/2; 6 x 9; 8 1/2 x 11

Bindings: [I] PB   [I] SS   [O] HC   [I] C/SB

Capabilities: [ ] Magazines   [ ] Galley Copies
               [ ] Journals   [ ] Demand Printing
               [X] Cookbooks   [X] 4-color Juvenile Books
               [X] Yearbooks   [ ] Annual Reports/Brochures
               [X] Catalogs   [X] Other Commercial Printing

Services: [ ] Typesetting   [ ] Teletypesetting
           [ ] Design and Pasteup   [ ] Editing
           [X] 4-color Printing   [X] Warehousing/Shipping

Terms: Net 30 days.

Semline, a member of the Book Manufacturers Institute and one of the 100 largest printers in the United States (with annual sales of $25,000,000), has a fully-automatic mechanical bindery. They also provide kit assembly and distribution services.

Service Printing Company
2725 Miller Street
San Leandro   CA 94577

415-352-7890
Jerry Edelstein
President

Quantities: Min: 1000   Max: 25,000   Opt: 5000

Book Sizes: 5 1/2 x 8 1/2; 7 x 9; 8 1/2 x 11; 11 x 17

Bindings: [X] PB   [X] SS   [X] HC   [X] C/SB

Capabilities: [ ] Magazines   [ ] Galley Copies
                [ ] Journals   [ ] Demand Printing
                [ ] Cookbooks   [ ] 4-color Juvenile Books
                [ ] Yearbooks   [ ] Annual Reports/Brochures
                [X] Catalogs   [X] Other Commercial Printing

Services: [ ] Typesetting   [ ] Teletypesetting
            [ ] Design and Pasteup   [ ] Editing
            [ ] 4-color Printing   [X] Warehousing/Shipping

Terms: Net 30 days.

In business since 1925, they specialize in producing training
and technical manuals for the computer/electronics industry.
Their routine delivery is 5 to 7 working days from receipt of
camera-ready copy.

Sexton Printing
250 E. Lothenbach
Saint Paul MN 55118

612-457-9255
James P. Sexton
Sales Manager

Quantities: Min: 10   Max: 20,000   Opt: 5000

Book Sizes: almost any size, including odd sizes

Bindings: [O] PB   [I] SS   [O] HC   [ ] C/SB

Capabilities: [X] Magazines   [ ] Galley Copies
                [X] Journals   [ ] Demand Printing
                [ ] Cookbooks   [ ] 4-color Juvenile Books
                [ ] Yearbooks   [X] Annual Reports/Brochures
                [X] Catalogs   [X] Other Commercial Printing

Services: [X] Typesetting   [X] Teletypesetting
            [X] Design and Pasteup   [ ] Editing
            [X] 4-color Printing   [X] Warehousing/Shipping

More details on the next page.

Sexton Printing continued

Terms: Net 30 days.

A sheetfed printer since 1949, Sexton specializes in printing publications (newsletters, house organs, magazines) for clubs, associations, and businesses. They have a sophisticated typesetting system that allows them to provide page proofs within two days after receiving already keyboarded material (either via disc or modem). They currently typeset, layout, print, bind, and provide fulfillment for over 150 publications, many in full-color.

The Sheridan Press   800-352-2210 / 717-632-3535
Fame Avenue   Paul T. Lydon
Hanover PA 17331   Vice President Sales & Marketing

Quantities: Min: 500   Max: 50,000   Opt: 10,000

Book Sizes: 5 1/2 x 8 1/2; 6 x 9; 6 7/8 x 10; 8 1/2 x 11

Bindings: [I] PB   [I] SS   [O] HC   [O] C/SB

Capabilities: [X] Magazines   [ ] Galley Copies
                [X] Journals   [ ] Demand Printing
                [X] Cookbooks   [ ] 4-color Juvenile Books
                [ ] Yearbooks   [ ] Annual Reports/Brochures
                [X] Catalogs   [ ] Other Commercial Printing

Services: [X] Typesetting   [X] Teletypesetting
            [X] Design and Pasteup [X] Editing
            [X] 4-color Printing   [X] Warehousing/Shipping

Terms: Net 30 days.

Sheriden, in business since 1915, specializes in printing, binding, and mailing high-quality publications, including medical and scientific journals and magazines. They also offer a very easy to use article reprint service. In short, they provide complete publishing services from design and redaction to subscription fulfillment and mailing list maintenance. They are noted for their excellent quality, dependable service, and on-time delivery.

```
* * * * * * * * * * * * * * * * * * * * * * * * * * * * * * *
*  Tip:  Stay away from tight registrations whenever possible.  *
*        They require extra set-up time and can cause problems. *
* * * * * * * * * * * * * * * * * * * * * * * * * * * * * * *
```

Skillful Means Press
1241 - 21st Street
Oakland CA 94607

415-839-3931
Alicia Fazio
Sales Representative

Quantities: Min: 2000   Max: 50,000   Opt: 20,000

Book Sizes: 5 1/2 x 8 1/2; 6 x 9; 8 1/2 x 11

Bindings: [X] PB   [X] SS   [X] HC   [X] C/SB

Capabilities: [X] Magazines   [ ] Galley Copies
               [X] Journals   [ ] Demand Printing
               [X] Cookbooks   [X] 4-color Juvenile Books
               [ ] Yearbooks   [X] Annual Reports/Brochures
               [X] Catalogs   [X] Other Commercial Printing

Services: [X] Typesetting   [ ] Teletypesetting
           [ ] Design and Pasteup   [X] Editing
           [X] 4-color Printing   [ ] Warehousing/Shipping

Terms: 50% down, net 30 days.

Skillful Means, formerly known as Dharma Press, has been in business for 15 years. "We are economical on brochures from 5000 to 200,000, booklets from 3000 to 100,000, and 4-color work from 1500 to 55,000." They meet their delivery schedules.

Snohomish Publishing
114 Avenue "C" / P O Box 499
Snohomish WA 98290

206-568-4121
Jeff Wise
Estimator

SP

Quantities: Min: 500   Max: not stated   Opt: 1000 - 5000

Book Sizes: 5 1/2 x 8 1/2; 8 1/2 x 11

Bindings: [I] PB   [I] SS   [O] HC   [O] C/SB

Capabilities: [X] Magazines   [ ] Galley Copies
               [ ] Journals   [ ] Demand Printing
               [X] Cookbooks   [ ] 4-color Juvenile Books
               [ ] Yearbooks   [ ] Annual Reports/Brochures
               [X] Catalogs   [X] Other Commercial Printing

Services: [X] Typesetting   [X] Teletypesetting
           [X] Design and Pasteup   [ ] Editing
           [X] 4-color Printing   [ ] Warehousing/Shipping

More details on the next page.

Snohomish Publishing continued

Terms: Negotiable.

Snohomish appears to be a smaller printing firm who specializes in working with first time and experienced self-publishers.
Judging from the user comment below, their prices are high for
ultra-short runs.

User Comment: "Not competitive under 3000 copies."

RATINGS	1	2	3	4	5	6	7	8	9	10	Ave	
Speed	–	–	–	1	–	–	–	–	–	–	---	1
Price	–	–	–	–	–	1	–	–	–	–	---	
Dependability	–	–	–	–	–	–	–	1	–	–	---	
Service . . .	–	–	–	–	–	–	–	1	–	–	---	
Quality . . .	–	–	–	–	–	–	–	1	–	–	---	
Overall . . .	–	–	–	–	–	–	1	–	–	–	---	

Southam Printing Ltd.   416-741-9700
2973 Weston Road, Box 510   Attn: Sales Representative
Weston, Ontario
M9N 3R3   Canada

U. S. customers may call 212-696-2083 for quotations.

Southam specializes in printing 4-color magazines, catalogs,
books, and annual reports. Apparently, though, they do only
larger runs. They did not respond to our recent printer survey.

Southeastern Printing Company   305-287-2141
3601 SE Dixie Hwy / P O 2476   Robert Anthony
Stuart FL 33495   Vice President Sales & Marketing

Florida customers may call their in-state toll-free WATS number: 800-432-8221. They also have a New York sales office; call
212-586-6953.

Quantities: Min: none   Max: none   Opt: not stated

Book Sizes: 5 1/2 x 8 1/2; 6 x 9; 8 1/2 x 11; 9 x 12

Bindings: [I] PB   [I] SS   [ ] HC   [I] C/SB

Capabilities: [X] Magazines   [ ] Galley Copies
               [X] Journals   [X] Demand Printing
               [X] Cookbooks   [X] 4-color Juvenile Books
               [ ] Yearbooks   [X] Annual Reports/Brochures
               [X] Catalogs   [X] Other Commercial Printing

Services: [X] Typesetting   [ ] Teletypesetting
           [X] Design and Pasteup [ ] Editing
           [X] 4-color Printing   [X] Warehousing/Shipping

Terms: 30 days with approved credit.

Southeastern has been in business for over 60 years as a general commercial sheet-fed printer. Besides books, periodicals, catalogs, and annual reports, they also print one to four-color art prints, greeting cards, brochures, folders, point-of-purchase displays, and packaging.

Southern Tennessee Publishing   615-722-5404   SP
P O Box 91   Paula Christian
Waynesboro   TN 38485   President

Quantities: Min: 50   Max: 2000   Opt: 500

Book Sizes: 4 x 7; 5 1/2 x 8 1/2; 8 1/2 x 11

Bindings: [O] PB   [I] SS   [O] HC   [O] C/SB

Capabilities: [ ] Magazines   [X] Galley Copies
               [X] Journals   [ ] Demand Printing
               [ ] Cookbooks   [ ] 4-color Juvenile Books
               [ ] Yearbooks   [ ] Annual Reports/Brochures
               [ ] Catalogs   [X] Other Commercial Printing

Services: [X] Typesetting   [ ] Teletypesetting
           [X] Design and Pasteup [X] Editing
           [X] 4-color Printing   [X] Warehousing/Shipping

Terms: Typeset books: 50% down, 50% with returned proofs.
        Camera-ready books: payment in full with order.
        Discounts on volume business.

Southern Tennessee is a small "but quick" (usually three weeks to finished books) printer serving self-publishers. They offer editing and proofreading done by English professionals. Their prices include free shipping, free copyright service, free cover design, and two free illustrations. Send for their standard price list. They give personal attention to all their customers.

Sowers Printing Company
N 10th & Scull St / PO Box 479
Lebanon PA 17042-0479

717-272-6667
Roy C. Bliley
Vice President Sales

Pennsylvania customers may call toll-free 800-692-7321; some East Coast customers may call toll-free 800-233-7028.

Sowers asked to be listed in this <u>Directory</u>, but they didn't return the printer survey in time for us to present a complete description. Here's what we know:

Quantities: unknown

Book Sizes: unknown, probably any standard size

Bindings: [O] PB   [I] SS   [ ] HC   [I] C/SB

Capabilities: [X] Magazines   [ ] Galley Copies
               [ ] Journals   [ ] Demand Printing
               [ ] Cookbooks   [ ] 4-color Juvenile Books
               [ ] Yearbooks   [X] Annual Reports/Brochures
               [X] Catalogs   [X] Other Commercial Printing

Services: [ ] Typesetting   [ ] Teletypesetting
           [X] Design and Pasteup [ ] Editing
           [X] 4-color Printing   [X] Warehousing/Shipping

Other Services: addressing and mailing services, including computer list maintenance. Overnight delivery delivery to most mid-Atlantic states.

Terms: Not stated.

Spilman Printing Company
1801 Ninth Street
Sacramento CA 95814

916-448-3511 / 800-448-3511 CA
Larry J. Burnette
Vice President of Operations

Quantities: Min: 1000   Max: 100,000   Opt: 15,000 - 25,000

Book Sizes: 5 1/2 x 8 1/2; 6 x 9; 7 x 10; 8 1/2 x 11

Bindings: [I] PB   [I] SS   [O] HC   [O] C/SB

Capabilities: [ ] Magazines   [ ] Galley Copies
               [X] Journals   [ ] Demand Printing
               [ ] Cookbooks   [ ] 4-color Juvenile Books
               [ ] Yearbooks   [X] Annual Reports/Brochures
               [X] Catalogs   [ ] Other Commercial Printing

Services: [ ] Typesetting   [ ] Teletypesetting
           [X] Design and Pasteup [ ] Editing
           [X] 4-color Printing   [ ] Warehousing/Shipping

Terms: Net 30 with approved credit.

Calling itself the "West Coast short-run book specialist," Spilman is beginning to establish a reputation as a reliable and quality book printer. Their prices tend to be lower than Delta Lithographs. They publish a quarterly newsletter, the Bookmaker, which is full of many worthwhile features. Send for a copy.

Note: 4-color printing and medium-sized (6 x 9 and 7 x 10) books are only available on their sheet-fed presses in runs of 15,000 or less.

Spilman is also capable of printing 3-D effects (for use with those funny green and red glasses.

User Comment: "Good printer, local for us, suits our needs for short-run books."

RATINGS	1	2	3	4	5	6	7	8	9	10	Ave	
Speed	-	-	-	-	-	-	-	1	-	-	---	1
Price	-	-	-	-	-	-	-	-	1	-	---	
Dependability	-	-	-	-	-	-	1	-	-	-	---	
Service . . .	-	-	-	-	-	-	-	1	-	-	---	
Quality . . .	-	-	-	-	-	1	-	-	-	-	---	
Overall . . .	-	-	-	-	-	-	-	1	-	-	---	

Staked Plains Press   806-655-1061
P O Box 816   Troy F. Martin
Canyon TX 79015   President

According to their listing in LMP, they are capable of doing short runs. However, they did not answer our printer surveys. They seem to be essentially a typesetter and printing broker with limited binding capabilities (spiral and comb binding).

```
* * * * * * * * * * * * * * * * * * * * * * * * * * * * * * *
*  Tip:  Use bleeds only when necessary.  Bleed pages often  *
*        need to be put on larger sheets of paper (to allow  *
*        for trimming), and the larger sheets will cost you  *
*        more money.  Also, bleed pages often require extra  *
*        set up and finishing labor -- costing more money.   *
* * * * * * * * * * * * * * * * * * * * * * * * * * * * * * *
```

Standard Printing Service   312-346-0499
162 N State Street   Attn: President
Chicago IL 60601

They also have not answered our printer surveys. Apparently
they are not interested in doing short-run book production,
although they offer both sheetfed and web printing.

==

Straus Printing Company   608-251-3222
1028 E Washington Av / PO 2118   Reed Jones
Madison WI 53701   Customer Service Representative

Quantities: Min: 5000   Max: 200,000   Opt: 15,000 - 50,000

Book Sizes: almost any size

Bindings: [I] PB   [I] SS   [O] HC   [O] C/SB

Capabilities: [X] Magazines   [ ] Galley Copies
                [X] Journals   [ ] Demand Printing
                [ ] Cookbooks   [ ] 4-color Juvenile Books
                [ ] Yearbooks   [X] Annual Reports/Brochures
                [X] Catalogs   [X] Other Commercial Printing

Services: [ ] Typesetting   [ ] Teletypesetting
            [X] Design and Pasteup   [ ] Editing
            [X] 4-color Printing   [X] Warehousing/Shipping

Terms: Net 30 with approved credit.

Their speciality is medium runs of high quality 4-color bound
materials (catalogs, annual reports, brochures), but they also
print posters up to 25" x 38" and other 4-color items. They can
provide perfectbinding in quantities from 5000 to 50,000 and saddlestitching in quantities from 15,000 to 200,000.   Other
bindings can be subcontracted to outside vendors.

==

The Studley Press   413-686-0441
151 E Housatonic Street   Thomas Reardon
Dalton MA 01226   President

Quantities: Min: 1000   Max: 20,000   Opt: 10,000

Book Sizes: 5 1/2 x 8 1/2; 6 x 9; 8 1/2 x 11

Bindings: [X] PB   [X] SS   [X] HC   [X] C/SB

Capabilities: [X] Magazines   [ ] Galley Copies
               [X] Journals   [ ] Demand Printing
               [ ] Cookbooks   [ ] 4-color Juvenile Books
               [ ] Yearbooks   [X] Annual Reports/Brochures
               [ ] Catalogs   [X] Other Commercial Printing

Services: [X] Typesetting   [X] Teletypesetting
           [X] Design and Pasteup [ ] Editing
           [X] 4-color Printing   [ ] Warehousing/Shipping

Terms: Net 30 days.

They emphasize their quality reproduction of photos and other halftone artwork. Note that their optimum print run is 10,000 or more copies.

Sweet Printing   512-255-1055
P O Box 49290   Dick Cluster
Austin   TX 78765   Estimator

A full service commercial printer, Sweet offers "complete book manufacturing" but apparently is not interested in doing short runs since they did not answer our printer survey.

John S. Swift Company   314-991-4300
1248 Research Blvd / P O 28252   Ben Heckel
St Louis   MO 63132   General Manager

A general commercial printer, Swift has three other printing plants:

1) 17 N. Loomis St., P. O. 7261, Chicago, IL 60607; 312-666-7020;
2) 2524 Spring Grove, Cincinnati, OH 45214; 513-721-4147; and
3) U.S. Route 46, Teterboro, NJ 07608; 201-288-2050.
   Swift also has sales offices in the following cities:

216-861-7070 Cleveland OH   414-276-6170 Milwaukee WI
816-531-5544 Kansas City MO

Quantities: Min: 100   Max: 15,000   Opt: 5000

For more details on their services, see the next page.

169

John S. Swift Company continued

Book Sizes: almost any size

Bindings: [I] PB [I] SS [ ] HC [I] C/SB

Capabilities: [ ] Magazines   [ ] Galley Copies
               [ ] Journals   [X] Demand Printing
               [ ] Cookbooks   [ ] 4-color Juvenile Books
               [ ] Yearbooks   [X] Annual Reports/Brochures
               [X] Catalogs   [X] Other Commercial Printing

Services: [X] Typesetting   [ ] Teletypesetting
           [X] Design and Pasteup   [ ] Editing
           [X] 4-color Printing   [ ] Warehousing/Shipping

Terms: Net 30 with approved credit.

In business since 1912, Swift is a general commercial printer who can print catalogs, prices lists, posters, manuals, office forms, brochures, maps and multi-color productions besides books.

Synthex Press   415-824-8282
2590 Folsum   Attn: Sales Representative
San Francisco CA 94110

They originally asked to be listed in this <u>Directory</u>, but then decided that they did not want to be listed as short-run book printers.

Taylor Publishing Company   214-637-2800
1550 W Mockingbird Lane   Randolph B. Marston
Dallas TX 75235   President

One of the 25 largest printers in the U.S. as well as a book publisher themselves, they have not answered our printer surveys. Apparently they do not do short runs. Perhaps their subsidiary, Newsfoto Publishing Company offers short-run capability -- they do multicolor sheetfed offset printing and color separations.

*** Use this <u>Directory</u> to help you find those book printers capable of fulfilling your changing printing and binding requirements for each new book you publish.

170

Telegraph Press
P O Box 1831
Harrisburg PA 17105

717-234-5091
Harold Baucum
Marketing Manager

A subsidiary of Commonwealth Communications Services, Telegraph has been in business since 1831 printing magazines, tabloids, and books. They also provide mailing and newsstand distribution services. However, they probably do not do short runs since they did not answer our printer surveys.

TGI Graphics
10 Lucon Drive
Deer Park NY 11729

516-586-1973
Morry Gropper
Sales Manager

They are a web offset printer not capable of doing short runs.

Thomson-Shore Inc.
7300 W Joy Road / P O Box 305
Dexter MI 48130-0305

313-426-3939
Ned Thomson
President

Quantities: Min: 35   Max: 7500   Opt: 2000

Book Sizes: 5 x 7, 5 1/2 x 8 1/2, 6 x 9, 7 x 10, 8 1/2 x 11

Bindings: [I] PB   [I] SS   [O] HC   [I] C/SB

Capabilities:
- [ ] Magazines
- [X] Journals
- [X] Cookbooks
- [ ] Yearbooks
- [X] Catalogs

- [ ] Galley Copies
- [ ] Demand Printing
- [ ] 4-color Juvenile Books
- [ ] Annual Reports/Brochures
- [ ] Other Commercial Printing

Services:
- [ ] Typesetting
- [ ] Design and Pasteup
- [X] 4-color Printing

- [ ] Teletypesetting
- [ ] Editing
- [ ] Warehousing/Shipping

Terms: To be worked out jointly.

Thomson-Shore is one of the top short-run book printers in the business. They are noted for their high quality and excellent service. They tied for first in the 1981 Small Press Review book printer survey and won several AIGA awards in 1984 for book printing.

More details on the next page.

171

They have no outside salesmen; you deal direct with their home office staff who answer all quotes within 24 hours. Thomson-Shore publishes a superb quarterly newsletter, <u>Printer's Ink</u>, which includes many tips and suggestions. Ask to be put on their mailing list.

They have printed over 15,000 titles (books, journals, catalogs, manuals) over the past ten years. On catalogs and other saddlestitched items they can print and bind up to 35,000 copies. 4-color work is limited to covers and inserts up to 32 pages.

User Comments: "They're absolutely most dependable, especially on runs under 2500. I'd recommend them to virtually everyone. Prices on the same day!" ... "Solid company. Top quality. Slow. Interested in clients." ... "Excellent quality printing, the best I've seen. Nice sharp crisp jet black printing, like the original or better." ... "A class operation, #1 in customer communication, but slow and relatively expensive."

Ratings	1	2	3	4	5	6	7	8	9	10	Ave	
Speed	1	-	1	1	3	1	1	1	-	-	4.89	9
Price	-	-	-	1	1	2	-	2	2	1	7.22	
Dependability	-	1	-	-	1	-	-	2	3	2	7.78	
Service . . .	-	-	-	-	1	-	1	2	3	2	8.33	
Quality . . .	-	-	-	-	1	-	1	-	2	5	8.89	
Overall . . .	-	-	-	-	1	1	-	3	4	-	7.89	

Times Litho
Times Litho Printing
Forest Grove OR 97116

?
Paul McGilvra

These people were recommended by one of their customers, but they did not answer the printer survey we sent them so we cannot provide any other details regarding their capabilities.

Tompson & Rutter Inc.
Dunbar Hill Road / P O Box 297
Grantham NH 03753

603-863-4392
Frances T. Rutter
President

They are printing brokers who apparently specialize in longer runs since they did not answer our printer surveys.

Torch Publications
5353 Mission Center Rd #124
San Diego CA 92108

619-299-2111
Ron Becijos
Consultant

Quantities: Min: 1000   Max: 75,000   Opt: 5000 - 10,000

Book Sizes: 5 1/2 x 8 1/2; 6 x 9; 8 1/2 x 11; 9 x 12

Bindings: [X] PB   [X] SS   [X] HC   [X] C/SB

Capabilities: [X] Magazines   [ ] Galley Copies
                [ ] Journals   [ ] Demand Printing
                [X] Cookbooks   [ ] 4-color Juvenile Books
                [ ] Yearbooks   [X] Annual Reports/Brochures
                [X] Catalogs   [X] Other Commercial Printing

Services: [X] Typesetting   [ ] Teletypesetting
            [X] Design and Pasteup   [X] Editing
            [X] 4-color Printing   [X] Warehousing/Shipping

Terms: Typeset books: 1/3, 1/3, 1/3; camera-ready books: 50/50.

  Torch provides total book production for small presses and
self-published authors, including editing, writing and marketing
services (publicity, direct mail promotions, advertising, distribution placement, and special market sales).

===

Town House Press
28 Midway Road
Spring Valley NY 10977

914-425-2232
Alvin Schultzberg
President

Quantities: Min: 200   Max: 5000   Opt: 2000

Book Sizes: almost any size from 4 1/4 x 7 to 8 1/2 x 11

Bindings: [X] PB   [X] SS   [X] HC   [X] C/SB

Capabilities: [ ] Magazines   [ ] Galley Copies
                [X] Journals   [ ] Demand Printing
                [X] Cookbooks   [ ] 4-color Juvenile Books
                [X] Yearbooks   [ ] Annual Reports/Brochures
                [X] Catalogs   [ ] Other Commercial Printing

Services: [X] Typesetting   [ ] Teletypesetting
            [X] Design and Pasteup   [X] Editing
            [X] 4-color Printing   [ ] Warehousing/Shipping

More details on the next page, so turn the page.

Town House Press continued

Terms: To be arranged.

   Town House specializes in providing quality printing of short-run books and journals. They can print 4-color books (interiors as well as covers) in runs of as few as 1500 copies. Town House is a good printer who pays attention to details. It was recommended by Judith Appelbaum and Nancy Evans in their book, How to Get Happily Published.

==

Tracor Publications   512-929-2222
6500 Tracor Lane   Doug McBride
Austin TX 78725   Marketing Manager

Quantities: Min: 50   Max: 50,000   Opt: 10,000

Book Sizes: almost any size

Bindings: [O] PB   [I] SS   [O] HC   [I] C/SB

Capabilities: [X] Magazines   [X] Galley Copies
               [X] Journals   [ ] Demand Printing
               [X] Cookbooks   [X] 4-color Juvenile Books
               [ ] Yearbooks   [X] Annual Reports/Brochures
               [X] Catalogs   [X] Other Commercial Printing

Services: [X] Typesetting   [X] Teletypesetting
           [X] Design and Pasteup [X] Editing
           [X] 4-color Printing   [X] Warehousing/Shipping

Terms: 1/3 down, 1/3 on delivery, 1/3 net 30.

   Tracor offers full design services (including technical, commercial, or architectural drawings and slide production). They are commercial printers specializing in magazine production and fulfillment, annual reports, sales literature, and technical manuals. They are, in their own words, "high quality, service-oriented, flexible, attentive to detail, attentive to customer"

* *
* **Tip:** To save money, you can have a number of your photos   *
*   shot as halftones at one time. The photos must have *
*   uniform contrasts and be the same size (or be reduced *
*   by the same percentage). Then the halftones can be   *
*   stripped into the production negatives as usual.   *
* *

TSO General Corporation
44-02 11th Street
Long Island City NY 11101

718-784-9550
Donald R. Skahan
President

Quantities: Min: 500   Max: 25,000   Opt: 15,000

Book Sizes: the regular sizes and multiples of them

Bindings: [X] PB   [X] SS   [ ] HC   [X] C/SB

Capabilities: [ ] Magazines   [ ] Galley Copies
               [X] Journals   [ ] Demand Printing
               [ ] Cookbooks   [ ] 4-color Juvenile Books
               [ ] Yearbooks   [X] Annual Reports/Brochures
               [ ] Catalogs   [X] Other Commercial Printing

Services: [X] Typesetting   [ ] Teletypesetting
           [ ] Design and Pasteup   [ ] Editing
           [ ] 4-color Printing   [X] Warehousing/Shipping

Terms: Net 30 days.

TSO is a general commercial printer with a multi-shift, 6 day a week operation, so they can handle lot of work. Note that they are black-and-white specialists.

Tucker Printers
80 Rockwood Place
Rochester NY 14610

716-271-4570
Daniel Tucker

They only do print runs over 10,000; hence, they are not a short-run book printer. We list them because Huenefeld lists them in their book publishing resource list, and it will save you time to know that they do not do short-runs.

Twin City Printery
P O Box 890, Industrial Park
Lewiston ME 04240

207-784-9181
Michael E. Wiesner
Sales Manager

Quantities: Min: 250   Max: 100,000+   Opt: 25,000

Book Sizes: almost any size.

For more about their capabilities, see the next page.

Twin City Printery continued

Bindings: [I} PB   [I] SS   [O] HC   [I] C/SB

Capabilities: [X] Magazines   [X] Galley Copies
               [X] Journals   [ ] Demand Printing
               [X] Cookbooks   [X] 4-color Juvenile Books
               [ ] Yearbooks   [X] Annual Reports/Brochures
               [X] Catalogs   [X] Other Commercial Printing

Services: [X] Typesetting   [X] Teletypesetting
           [X] Design and Pasteup [ ] Editing
           [X] 4-color Printing   [X] Warehousing/Shipping

Terms: To be discussed.

   They are a full-service printing facility (also with a two-
shift operation) capable of doing almost everything in house
except for casebinding. Their offer their best prices on runs of
25,000 or more.

Unicorn Press   714-546-7320
2136 S Wright Street, Box 2278   Arthur Vanderee
Santa Ana CA 92705   President

Quantities: Min: 100   Max: 10,000   Opt: 5000

Book Sizes: very flexible, 5 1/2 x 8 1/2 to 8 1/2 x 11

Bindings: [I] PB   [I] SS   [O] HC   [I] C/SB

Capabilities: [X] Magazines   [X] Galley Copies
               [X] Journals   [ ] Demand Printing
               [X] Cookbooks   [ ] 4-color Juvenile Books
               [ ] Yearbooks   [X] Annual Reports/Brochures
               [X] Catalogs   [X] Other Commercial Printing

Services: [X] Typesetting   [X] Teletypesetting
           [X] Design and Pasteup [ ] Editing
           [X] 4-color Printing   [ ] Warehousing/Shipping

Terms: Not stated.

   A "quality shop" since 1963, they specialize in printing and
binding software and hardware manuals, newsletters, employee
booklets, brochures, and other business publications. Their
typesetting system can read 200 floppy disk formats. They also
can design and produce technical artwork.

User Comment: "Difficult to get specific breakdown of pricing (and expensive)."

RATINGS	1	2	3	4	5	6	7	8	9	10	Ave	
Speed	-	-	-	1	-	-	-	-	-	-	---	1
Price	-	1	-	-	-	-	-	-	-	-	---	
Dependability	-	-	-	-	1	-	-	-	-	-	---	
Service . . .	-	-	-	-	1	-	-	-	-	-	---	
Quality . . .	-	-	-	-	-	-	-	-	1	-	---	
Overall . . .	-	-	-	1	-	-	-	-	-	-	---	

United Color Press
240 W Fifth Street
Dayton OH 45402

513-461-5150
Dan Duffy
President

United specializes in trade magazines, catalogs, and directories but apparently not in runs under 10,000 copies. They did not answer our printer surveys.

Universal Printing Company
1701 Macklind Avenue
St Louis MO 63110

314-771-6900
Thomas Emerson
President

Although they offer both sheetfed and web offset printing as well as letterpress, they apparently do not do short runs. They also did not answer our printer surveys.

University Press
21 East Street
Winchester MA 01890

617-729-8000
Bertha Teel
President

A division of Publishers Book Bindery, they specialize in India paper editions. They may do short runs; however, we are not sure since they too did not answer our printer surveys.

* *
* **Tip:** Always ask for samples of the printer's work.   *
* *

University Printers
1120 St Joseph Avenue
Berrien Springs MI 49103

616-471-3236
Fuad Mashni
Sales Manager

There seems to be almost a curse on printers whose names begin with the letter "U" since here's another one who did not answer the printer survey we sent them. Oh, well, at least they have a good recommendation from one of their users (see below).

User Comment: "A good press with solid customer support. They do many jobs from the Chicago area. Perfect binding and saddlestitching in plant."

RATINGS	1	2	3	4	5	6	7	8	9	10	Ave	
Speed	-	-	-	-	1	-	-	-	-	-	---	1
Price	-	-	-	-	1	-	-	-	-	-	---	
Dependability	-	-	-	-	-	1	-	-	-	-	---	
Service . . .	-	-	-	-	-	-	1	-	-	-	---	
Quality . . .	-	-	-	-	1	-	-	-	-	-	---	
Overall . . .	-	-	-	-	1	-	-	-	-	-	---	

Van Volumes
15 Railroad Avenue
Wilbraham MA 01095

413-596-2113
Hermine Deso
President

Quantities: Min: 20   Max: 5000   Opt: 100 - 1000

Book Sizes: anything from 4 x 4 to 8 1/2 x 11

Bindings: [I] PB   [I] SS   [O] HC   [I] C/SB

Capabilities: [ ] Magazines   [X] Galley Copies
                 [X] Journals   [ ] Demand Printing
                 [X] Cookbooks   [ ] 4-color Juvenile Books
                 [ ] Yearbooks   [ ] Annual Reports/Brochures
                 [ ] Catalogs   [ ] Other Commercial Printing

Services: [ ] Typesetting   [ ] Teletypesetting
           [ ] Design and Pasteup [ ] Editing
           [ ] 4-color Printing   [ ] Warehousing/Shipping

Terms: 1/2 down until credit is established. 2% 10, net 30
        days thereafter.

A new company, Van Volumes specializes in ultra-short runs of galley proofs, journals, and reprints. They have a standard price chart, charging by the page.

They charge as little as one cent per page for 100 copies of a 5 1/2 x 8 1/2 book (hence, a 160-page book would cost $1.60 per book for 100 copies). Prices are higher for 6 x 9 or 8 1/2 x 11 books (1.8 cents per page for 100 copies). Prices are much less for 1000 copies (.55 cents per page for 5 1/2 x 8 1/2; .85 cents per page for 6 x 9 or 8 1/2 x 11).

Their standard price includes the costs of paper, cover stock (65 lb. Navajo in a choice of colors), printing, and perfect binding.

Their prices are low, but they are slow (40 working days).

RATINGS	1	2	3	4	5	6	7	8	9	10	Ave	
Speed	-	-	-	1	-	-	-	-	-	-	---	1
Price	-	-	-	-	-	-	-	-	-	1	---	
Dependability	-	-	-	1	-	-	-	-	-	-	---	
Service . . .	-	-	-	1	-	-	-	-	-	-	---	
Quality . . .	-	-	-	1	-	-	-	-	-	-	---	
Overall . . .	-	-	-	1	-	-	-	-	-	-	---	

Versa Press   309-822-8272
Spring Bay Road / RR 1   Tom Mulvaney
East Peoria IL 61611   Production Coordinator

Quantities: Min: 300   Max: 20,000   Opt: 5000 - 10,000

Book Sizes: 4 1/4 x 7; 5 1/2 x 8 1/4; 6 x 9; 8 1/2 x 11

Bindings: [I] PB   [I] SS   [O] HC   [O] C/SB

Capabilities: [ ] Magazines   [ ] Galley Copies
               [ ] Journals   [ ] Demand Printing
               [X] Cookbooks   [ ] 4-color Juvenile Books
               [ ] Yearbooks   [ ] Annual Reports/Brochures
               [X] Catalogs   [ ] Other Commercial Printing

Services: [ ] Typesetting   [ ] Teletypesetting
           [ ] Design and Pasteup   [ ] Editing
           [ ] 4-color Printing   [ ] Warehousing/Shipping

Terms: Net 30 days.

Versa provides a breakdown on costs with their quotes so you can tell exactly how much each step in the book printing process will cost you. Note that Versa can print mass-market sized books in small quantities.

Vicks Lithograph & Printing
P O Box 270
Yorkville NY 13495

315-736-9346
Dwight E. Vicks Jr.
President

Although they are listed in LMP as capable of doing short to
medium runs, they sent us a letter last year saying that they
don't print runs shorter than 10,000 copies. Apparently that is
still the case since they did not respond to our printers surveys
this year.

Victor Graphics
200 N Bentalou, P.O. Box 4446
Baltimore MD 21223

301-233-8300
Al Etherton
Estimator

Washington, DC customers may call 202-452-0802.

Quantities: Min: 500   Max: 25,000   Opt: 10,000

Book Sizes: 5 1/2 x 8 1/2; 6 x 9; 7 x 10; 8 1/2 x 11; 9 x 12

Bindings: [I] PB   [I] SS   [O] HC   [O] C/SB

Capabilities:
- [ ] Magazines
- [X] Journals
- [X] Cookbooks
- [ ] Yearbooks
- [X] Catalogs
- [ ] Galley Copies
- [ ] Demand Printing
- [X] 4-color Juvenile Books
- [X] Annual Reports/Brochures
- [X] Other Commercial Printing

Services:
- [X] Typesetting
- [ ] Design and Pasteup
- [X] 4-color Printing
- [ ] Teletypesetting
- [ ] Editing
- [ ] Warehousing/Shipping

Terms: Net 30.

Victor Graphics has taken over the business of Publication
Press (same address and phone number). Their prices are quite
reasonable; they sent us one of the lowest quotes we received for
the printing of this edition of the Directory. Their quality and
service are also supposed to be good.

RATINGS	1	2	3	4	5	6	7	8	9	10	Ave	
Speed	-	-	-	-	-	-	-	1	-	-	---	1
Price	-	-	-	-	-	-	-	-	-	1	---	
Dependability	-	-	-	-	-	-	-	1	-	-	---	
Service . . .	-	-	-	-	-	-	-	-	1	-	---	
Quality . . .	-	-	-	-	-	-	-	1	-	-	---	
Overall . . .	-	-	-	-	-	-	-	-	1	-	---	

Viking Press
7000 Washington Avenue S
Eden Prairie MN 55344

612-941-8780
Attn: Sales Representative

   Viking is a fairly new press which, we believe, specializes in
print runs over 10,000. We're not sure because they did not
respond to the printer survey we sent them. (They may have taken
over business from Colwell Press, which is not out of business.)

═══

Vogue Printers
2421 Green Bay Road
North Chicago IL 60064

312-689-4044
Pete Deperte
Sales Manager

Quantities: Min: 10   Max: 100,000   Opt: not stated

Book Sizes: almost any size

Bindings: [I] PB   [I] SS   [ ] HC   [I] C/SB

Capabilities: [ ] Magazines   [ ] Galley Copies
               [ ] Journals   [ ] Demand Printing
               [X] Cookbooks   [ ] 4-color Juvenile Books
               [ ] Yearbooks   [X] Annual Reports/Brochures
               [X] Catalogs   [X] Other Commercial Printing

Services: [X] Typesetting   [X] Teletypesetting
           [X] Design and Pasteup   [ ] Editing
           [X] 4-color Printing   [X] Warehousing/Shipping

Terms: Net 30.

   Vogue is a general commercial printer that also prints books
and manuals. Try them out if you are located in the Chicago
area. At least one user rates them very highly. See the
comments below:

   User Comment: "Super quality, willing to please, good fast
service. Doesn't specialize in book printing, but has always
done a good job on print runs of 500 or 1000 perfectbound."

RATINGS	1	2	3	4	5	6	7	8	9	10	Ave	
Speed	-	-	-	-	-	-	-	1	-	-	---	1
Price	-	-	-	-	-	-	1	-	-	-	---	
Dependability	-	-	-	-	-	-	-	-	-	1	---	
Service . . .	-	-	-	-	-	-	-	-	1	-	---	
Quality . . .	-	-	-	-	-	-	-	-	1	-	---	
Overall . . .	-	-	-	-	-	-	-	-	1	-	---	

Von Hoffman Press
1000 Camera Avenue
St Louis MO 63126

314-966-0909
Attn: Sales Representative

They have never answered any of our RFQ's or survey forms.
They apparently do not do short runs.

Waldon Press
216 W 18th Street
New York NY 10011

212-691-9220
William H. Donat
President

Waldon is a commercial/financial printer who also prints books
and periodicals in English or foreign languages. They supposedly
print any quantity from 3000 to 25,000, but they have never answered any of our RFQ's or printer surveys.

Wallace Press
4600 W Roosevelt Road
Hillside IL 60162

312-626-2000
Tom Franke
Sales Manager

A division of Wallace Computing Services, they apparently
don't do runs below 10,000 copies even though they have both
sheetfed and web offset capabilities. They did not answer our
printer survey so we cannot be sure.

Walsworth Publishing
306 N Kansas Avenue
Marceline MO 64658

816-376-3543
John Tucker, Manager
Commercial Book Division

Quantities: Min: 500   Max: 20,000   Opt: 5000 - 6000

Book Sizes: any combination up to and including 9 x 12

Bindings: [I] PB   [I] SS   [I] HC   [I] C/SB

Capabilities: [X] Magazines   [X] Galley Copies
                 [X] Journals   [X] Demand Printing
                 [X] Cookbooks   [X] 4-color Juvenile Books
                 [X] Yearbooks   [ ] Annual Reports/Brochures
                 [X] Catalogs   [X] Other Commercial Printing

Services: [X] Typesetting   [X] Teletypesetting
         [X] Design and Pasteup   [ ] Editing
         [X] 4-color Printing   [X] Warehousing/Shipping

Terms: Net 30 days with approved credit.

    Walsworth is the printer of this edition of the Directory; you
are holding a sample of their work in your hands right now.
But don't blame them for the typesetting (we did that via our
computer and a laser printer).

    They offer very competitive prices on 4 or 5-color casebound
books (for example, coffee-table books, colorful travel guides,
and fancy cookbooks). They also have a separate division that
prints school yearbooks and special military publications.

    Because they do school yearbooks, they have a large layout and
design department capable of working with anyone who has little
graphics or printing experience. They do excellent work. They
have printed thousands of titles, both hard and softcover, for
major publishers as well as smaller publishers.

RATINGS	1	2	3	4	5	6	7	8	9	10	Ave	
Speed	-	-	-	-	-	-	-	1	-	-	---	1
Price	-	-	-	-	-	-	-	-	-	1	---	
Dependability	-	-	-	-	-	-	-	-	1	-	---	
Service . . .	-	-	-	-	-	-	-	-	1	-	---	
Quality . . .	-	-	-	-	-	-	-	-	1	-	---	
Overall . . .	-	-	-	-	-	-	-	-	1	-	---	

Walter's Publishing   800-447-3274 / 507-835-3691
Route 3   Wayne J. Dankert
Waseca MN 56093   General Manager

Quantities: Min: 200   Max: 3000   Opt: 500 - 1000

Book Sizes: 7 1/4 x 8 3/4 cookbook specialists

Bindings: [ ] PB   [ ] SS   [ ] HC   [I] C/SB

Capabilities: [ ] Magazines   [ ] Galley Copies
               [ ] Journals   [ ] Demand Printing
               [X] Cookbooks   [ ] 4-color Juvenile Books
               [ ] Yearbooks   [ ] Annual Reports/Brochures
               [ ] Catalogs   [ ] Other Commercial Printing

    More details about Walter's Cookbooks on the next page.

Walter's Cookbooks continued

Services: [X] Typesetting   [ ] Teletypesetting
           [X] Design and Pasteup   [ ] Editing
           [ ] 4-color Printing   [ ] Warehousing/Shipping

Terms: 90 day finance terms with no downpayment.

Walters produces personalized standard-format cookbooks for fundraising organizations. They will also print free publicity releases for the organizations. Send for their free cookbook fundraising kit which details their standard prices and formats. Note that they offer excellent finance terms for fundraising organizations: no downpayment plus 90 days to pay after you receive your books.

═══

Waverly Press Inc.   301-528-4000
428 E. Preston St   Wilbert McNamara
Baltimore MD 21202   Sales Manager

Quantities: Min: 1000   Max: 100,000+   Opt: 5000 - 20,000

Book Sizes: 6 x 9; 7 x 10; 8 1/2 x 11

Bindings: [I] PB   [I] SS   [ ] HC   [ ] C/SB

Capabilities: [X] Magazines   [ ] Galley Copies
               [X] Journals   [ ] Demand Printing
               [ ] Cookbooks   [ ] 4-color Juvenile Books
               [ ] Yearbooks   [ ] Annual Reports/Brochures
               [ ] Catalogs   [ ] Other Commercial Printing

Services: [X] Typesetting   [X] Teletypesetting
           [X] Design and Pasteup   [X] Editing
           [X] 4-color Printing   [X] Warehousing/Shipping

Terms: Net 30 with approved credit.

In business since 1890, Waverly is one of the hundred largest printers in the United States with annual sales over $25,000,000. Their subsidiary, Williams & Wilkins, publishes books and magazines in the fields of health, medicine, and science.

Waverly offers a full range of services for any association or other publisher of periodicals. They specialize in print runs of 5000 to 20,000 and can do journals in runs as low as 3000. They offer fulfillment and list maintenance as well as typesetting, design, editing, printing, and binding.

The Webb Company
1999 Shepard Road
St Paul MN 55116

612-690-7200
Robert Sallman
President

Webb, with over $100,000,000 in annual sales, is one of the 25 largest printers in the United States. Besides books, the print many trade, farm, and flight magazines. However, they do not seek out short-run printing jobs. They have not answered our printers surveys.

Webcom Ltd.
3480 Pharmacy Avenue
Scarborough, Ontario
M1W 3G3   Canada

416-496-1000
Warren D. Wilkins
President

A member of the Book Manufacturers Institute, Webcom prints books, directories, and catalogs -- but only in longer runs. They also did not answer our recent printer surveys.

Webcrafters
2211 Fordem Ave / P O Box 7608
Madison WI 53707

800-356-8200 / 608-244-3561
W. Jerome Frautschi
Vice President Sales

Webcrafters is one of the 65 largest printers in the United States, with annual sales over $35,000,000. They have sent us a letter stating that they do not do runs under 10,000.

Fred Weidner & Sons Printers
111 Eighth Avenue
New York NY 10011

212-989-1070
Fred Weidner III
President

They are publishing consultants and printing brokers. They can handle design, editorial, typesetting, plus printing -- doing as little or as much as their customer requires.

```
* * * * * * * * * * * * * * * * * * * * * * * * * * * * * * * *
*  Tip:  In short runs, your most significant price breaks     *
*        come at 3000 copies or 5000 copies.                   *
* * * * * * * * * * * * * * * * * * * * * * * * * * * * * * * *
```

West Coast Print Center
1915 Essex Street
Berkeley CA 94703

415-849-2746
Attn: Sales Representative

Like their East Coast cousin (see the Print Center), the West
Coast Print Center does jobs primarily for literary and arts
publishers. We do not know their exact services and capabilities
because they did not respond to the printer survey we sent them.

West Side Graphics
711 Amsterdam Avenue
New York NY 10025

212-222-9304 / 212-666-3218
Ed McDermott
Owner

Quantities: Min: 1000   Max: 20,000   Opt: 10,000

Book Sizes: almost any size

Bindings: [X] PB   [X] SS   [X] HC   [ ] C/SB

Capabilities: [X] Magazines   [ ] Galley Copies
                [X] Journals   [ ] Demand Printing
                [ ] Cookbooks   [ ] 4-color Juvenile Books
                [ ] Yearbooks   [X] Annual Reports/Brochures
                [ ] Catalogs   [X] Other Commercial Printing

Services: [X] Typesetting   [ ] Teletypesetting
            [X] Design and Pasteup [X] Editing
            [ ] 4-color Printing   [X] Warehousing/Shipping

Terms: Net 60 days.

They offer the best terms of any printer listed in this book
(other than the fundraising cookbook specialists): net 60 days
with approved credit.

Westview Press
5500 Central Avenue
Boulder CO 80301

303-444-3541
Douglas Kubach
Vice-President

Quantities: Min: 50   Max: 2000   Opt: 200 - 1200

Book Sizes: 5 1/2 x 8 1/2; 6 x 9; 8 1/2 x 11; 8 3/4 x 11 3/4

Bindings: [I] PB   [I] SS   [O] HC   [O] C/SB

Capabilities: [X] Magazines   [X] Galley Copies
               [X] Journals   [X] Demand Printing
               [X] Cookbooks   [ ] 4-color Juvenile Books
               [ ] Yearbooks   [X] Annual Reports/Brochures
               [X] Catalogs   [X] Other Commercial Printing

Services: [X] Typesetting   [X] Teletypesetting
            [X] Design and Pasteup [X] Editing
            [ ] 4-color Printing   [X] Warehousing/Shipping

Terms: Net 30 to net 60.

Associated with its own publishing house (a leading scholarly publisher), Westview has been in business since 1975 as an ultra-short run book specialist. They typeset, print, and distribute scholarly publications (both their own and others). Their typesetting system can handle five different methods for input: handwritten manuscript, optical scanner, mag tape, floppy disc, and telecommunications.

White Arts Inc.   317-638-3564 / 800-382-5022 IN
1203 E Saint Clair Street   Garry Hiott
Indianapolis IN 46202   Estimator

They have four printing plants in Indiana -- one in Spencer, one in Lafayette, and two in Indianapolis. Indiana customers may call their in-state toll-free number for quotations and service.

Quantities: Min: 100   Max: 15,000   Opt: 5000 - 10,000

Book Sizes: 5 1/2 x 8 1/2; 6 x 9; 8 1/2 x 11

Bindings: [O] PB   [I] SS   [O] HC   [I] C/SB

Capabilities: [X] Magazines   [ ] Galley Copies
               [X] Journals   [X] Demand Printing
               [ ] Cookbooks   [ ] 4-color Juvenile Books
               [ ] Yearbooks   [X] Annual Reports/Brochures
               [X] Catalogs   [X] Other Commercial Printing

Services: [X] Typesetting   [X] Teletypesetting
            [X] Design and Pasteup [ ] Editing
            [X] 4-color Printing   [X] Warehousing/Shipping

Terms: Net 30 with approved credit.

For comments on their capabilities and services, please turn the page.

White Arts Inc. continued

In business for over 50 years, White Arts is a general commercial printer of catalogs, manuals, journals, and reports. They can also print such speciality items as computer cards, bridge tickets, and payroll checks. They offer saddlestitching and comb binding in-house; other bindings can be subcontracted to outside vendors.

Whitehall Printing   312-541-9290
1200 S Willis   Mike Hirsch
Wheeling IL 60090   President

Quantities: Min: 500   Max: 5000   Opt: not stated

Book Sizes: 5 1/2 x 8 1/2; 6 x 9; 8 1/2 x 11

Bindings: [X] PB   [X] SS   [ ] HC   [ ] C/SB

Capabilities: [ ] Magazines   [ ] Galley Copies
               [ ] Journals   [ ] Demand Printing
               [ ] Cookbooks   [ ] 4-color Juvenile Books
               [ ] Yearbooks   [ ] Annual Reports/Brochures
               [X] Catalogs   [ ] Other Commercial Printing

Services: [ ] Typesetting   [ ] Teletypesetting
           [ ] Design and Pasteup [ ] Editing
           [ ] 4-color Printing   [ ] Warehousing/Shipping

Terms: Not stated.

Whitehall may be the least expensive book printer in the United States (example: 2000 copies of a 196-page perfectbound book for $1500.00). However, they do not reproduce halftones well and have rejected jobs which require very fine printing. They are known to be very slow.

The quality of their work has gotten worse as their workload has increased. We would recommend using them only when you have a book which does not require very high quality, and when you want to get a rock-bottom price.

User Comments: "Their reputation has gone downhill lately. Their quality is not as good as it once was." ... "I had to drop them -- bad service and quality." ... "Slow (60 days). Good quality. 50# paper only. Best prices in the U.S." ... "Poor on some halftones but will use them again."

more comments on the next page

188

"Although price is cheapest, print quality is average to below average, and they are hard to reach. They never get back to you when you leave a message." ... "They are slow."

RATINGS	1	2	3	4	5	6	7	8	9	10	Ave	
Speed	1	-	1	-	2	-	-	-	-	-	3.5	4
Price	-	-	-	-	-	-	-	-	1	3	9.75	
Dependability	-	1	1	-	1	-	-	1	-	-	4.5	
Service . . .	-	1	1	1	-	-	-	1	-	-	4.25	
Quality . . .	-	-	-	2	-	-	1	1	-	-	5.75	
Overall . . .	-	1	-	1	1	-	-	1	-	-	4.75	

Whittet & Shepperson Printers   804-649-9047
Third & Canal / P O Box 553   Steven Fitchett
Richmond VA 23204   Vice President Sales

They advertise that they give special attention to private, limited edition books. Their prices, however, for saddlestitch books is high. Since they have not answered our last two printer surveys, we are unable to give you more details regarding their services and capabilities.

Wickersham Printing Company   717-299-5731 / 212-925-7550
2959 Old Tree Drive   Arthur Dean Jr.
Lancaster PA 17603-4080   Vice President Sales

Quantities: Min: 500   Max: 200,000   Opt: 5000

Book Sizes: 5 x 7; 5 1/2 x 8 1/2; 6 x 9; 7 x 10; 8 1/2 x 11

Bindings: [I] PB   [I] SS   [O] HC   [O] C/SB

Capabilities: [ ] Magazines   [ ] Galley Copies
          [X] Journals   [ ] Demand Printing
          [X] Cookbooks   [ ] 4-color Juvenile Books
          [ ] Yearbooks   [ ] Annual Reports/Brochures
          [X] Catalogs   [ ] Other Commercial Printing

Services: [X] Typesetting   [ ] Teletypesetting
       [ ] Design and Pasteup   [ ] Editing
       [ ] 4-color Printing   [X] Warehousing/Shipping

More information on the next page.

Wickersham Printing Company continued

Terms: Negotiable.

Full service printers for over 123 years, Wickersham can print both one or two color books, journals, catalogs, and cookbooks. Note that they can provide oblong binding of the regular sizes (that is, they can bind the short side or the long side of a book).

Willcox Press
P O Box 9
Ithaca NY 14850

607-272-1212
J. Kevin Fahy
Marketing Manager

Willcox Press is an offset printer with perfectbound, saddle-stitch, and spiral bound capabilities. They are brokers for complete book manufacturing. However, they probably do not work with short runs since they did not reply to our printer surveys.

Wood & Jones
139 W Colorado Boulevard
Pasadena CA 91105-1983

213-681-9663
Richard B. Wood
President

Quantities: Min: 100   Max: 25,000   Opt: 5000 - 10,000

Book Sizes: almost any size

Bindings: [O] PB   [I] SS   [O] HC   [I] C/SB

Capabilities: [ ] Magazines   [X] Galley Copies
                 [X] Journals   [ ] Demand Printing
                 [ ] Cookbooks   [ ] 4-color Juvenile Books
                 [ ] Yearbooks   [X] Annual Reports/Brochures
                 [X] Catalogs   [X] Other Commercial Printing

Services: [X] Typesetting   [ ] Teletypesetting
             [X] Design and Pasteup   [ ] Editing
             [X] 4-color Printing   [ ] Warehousing/Shipping

Terms: Net 30 with approved credit.

Wood & Jones is a general commercial printer who typesets and prints many technical manuals. Their typesetting department can accept input from both computer and word processor diskettes.

Worzalla Publishing Company
3535 Jefferson Street
Stevens Point WI 54481

715-344-9600
David Varney
Senior Vice President

They have sales offices in the following places:

404-961-9323 Atlanta GA
617-749-1188 Boston MA
312-726-1555 Chicago IL

414-276-3378 Milwaukee WI
212-986-2941 New York NY

Quantities: Min: 1000   Max: 100,000   Opt: 5000 - 25,000

Book Sizes: almost any size, but they specialize in large size
(8 1/2 x 11) and odd size books

Bindings: [I] PB   [I] SS   [I] HC   [I] C/SB

Capabilities: [X] Magazines   [ ] Galley Copies
               [X] Journals   [ ] Demand Printing
               [X] Cookbooks   [X] 4-color Juvenile Books
               [ ] Yearbooks   [ ] Annual Reports/Brochures
               [X] Catalogs   [ ] Other Commercial Printing

Services: [X] Typesetting   [ ] Teletypesetting
           [ ] Design and Pasteup [ ] Editing
           [X] 4-color Printing   [X] Warehousing/Shipping

Terms: Flexible.

In business since 1898, Worzalla can print almost any size of
book including many odd sizes. Their speciality is full-color
illustrated children's books. They offer more competitive prices
when doing medium runs of 15,000 to 30,000. The quality of their
work is very good.

User Comments: "Excellent sales help." ... "High quality on
halftones, but expensive."

RATINGS	1	2	3	4	5	6	7	8	9	10	Ave	
Speed	-	-	-	-	1	-	-	-	1	-	---	2
Price	-	1	-	-	-	-	-	-	1	-	---	
Dependability	-	-	-	-	-	-	1	-	1	-	---	
Service	-	-	-	-	1	-	-	-	-	1	---	
Quality	-	-	-	-	-	-	-	1	1	-	---	
Overall	-	-	-	-	-	-	1	-	1	-	---	

```
* * * * * * * * * * * * * * * * * * * * * * * * * * * * * * * *
*  Tip:  Use photographs only when necessary since each half-  *
*        tone can add anywhere from $5 to $25 in prep costs.    *
* * * * * * * * * * * * * * * * * * * * * * * * * * * * * * * *
```

Xerographic Reproduction Center 201-871-4011
400 S Dean Street Sanford Saunders
Englewood Cliff NJ 07631 Sales Representative

A specialist in short runs of reader's copies and journal
reprints, they offer very fast service (4 to 7 working days, or
48 hours if a real emergency). Since they did not respond to our
printer survey, we can give no details about their services.

Yarrow Inc. 201-273-8390
P O Box 442 George Haralambous
Summit NJ 07901 Sales Representative

A sheetfed as well as web printer, they can print 4-color
books and other bound publications. They can also add fragrance.
Apparently, though, they do not do short runs; they did not
answer our recent printer survey.

William Yates / Printer ?
P O Box 237 William Yates
Ozark MO 65721 Owner

Since he was used by two of the publishers responding to our
user survey, we list this printer. We do not recommend him for
short-run work. He can only provide taped side-stiched binding,
and the quality of his work is not good. Indeed, it is possible
that he is no longer in business (we received no response to the
printer survey we sent him).

User Comments: "Stapled and taped bindings only." ... "His
printing is average, but 1/4 of the books were damaged in
shipping. When we asked him to reprint the damaged books, the
job was so sloppy we could not use it (the mat was put on the
press crooked). We would never use or recommend him if you want
saleable work."

RATINGS	1	2	3	4	5	6	7	8	9	10	Ave	
Speed	-	-	1	-	-	-	-	-	-	-	---	2
Price	-	-	-	-	-	-	-	1	-	-	---	
Dependability	-	-	-	-	-	-	1	-	-	-	---	
Service . . .	-	-	-	-	-	-	-	1	-	-	---	
Quality . . .	-	-	-	-	-	-	1	-	-	-	---	
Overall . . .	1	-	-	-	-	-	1	-	-	-	---	

WALSWORTH
PRESS
COMPANY
INC.

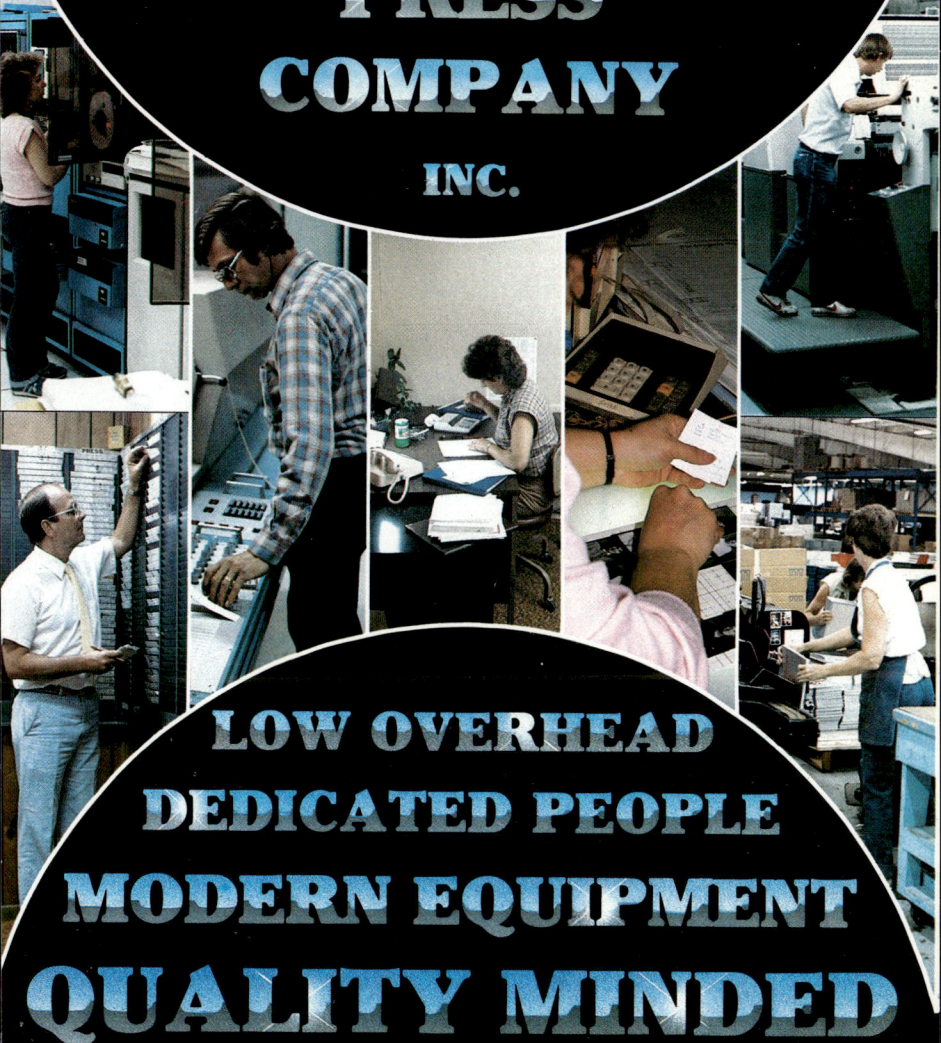

LOW OVERHEAD
DEDICATED PEOPLE
MODERN EQUIPMENT
QUALITY MINDED

you are not getting the quality and service you deserve from your present printer then give us a chance to quote your next book. Call or write:

WALSWORTH PRESS COMPANY, INC. • 306 NORTH KANSAS • MARCELINE, MO 64658

(816) 376-3543 • Ask for JOHN TUCKER or MARK ANDERSON

THIS CHILD ABUSER NEVER LAID A HAND ON ANYONE.

He screams and yells. Words are his weapons.

And the things he says can do as much damage as a punch in the mouth.

Verbally abused children who constantly hear, "You're no good. You're a lousy kid" grow up believing it.

They develop a negative self-image and very often become abusive parents themselves, never quite realizing that they have a problem. Or that they can change the way they are.

That's where Parents Anonymous comes in.

Parents Anonymous is a self-help organization designed to help parents handle the responsibilities of being parents. And to help them keep their frustrations from taking the form of child abuse.

At Parents Anonymous meetings parents share their feelings and experiences with others who have the same problems.

If you know someone who needs our help, or if you need someone to talk with, call your local chapter of Parents Anonymous or the Local Hotline 1-800-782-3320.

You'll find a quiet voice at the other end.

PARENTSANONYMOUS
We know it's tough being a parent!
National 1-800-421-0353
Kansas 1-800-332-6378

Yoder's Printing
RR #2, Box 39
Middleburg IN 46540

no phone
Fred O. Yoder
Owner

Quantities: Min: 50   Max: open   Opt: 10,000

Book Sizes: 5 1/2 x 8 1/2; 6 x 9; 8 1/2 x 11

Bindings: [ ] PB   [I] SS   [ ] HC   [I] C/SB

Capabilities: [ ] Magazines   [ ] Galley Copies
                [ ] Journals   [ ] Demand Printing
                [ ] Cookbooks   [ ] 4-color Juvenile Books
                [ ] Yearbooks   [X] Annual Reports/Brochures
                [ ] Catalogs   [ ] Other Commercial Printing

Services: [ ] Typesetting   [ ] Teletypesetting
            [X] Design and Pasteup   [ ] Editing
            [ ] 4-color Printing   [ ] Warehousing/Shipping

Terms: Net 30 days.

A new printer, they specialize in directories, pamphlets, and booklets. They offer "very reasonable prices if in larger quantities." Apparently, though, they are still ironing out their procedures -- their speed, service, and quality are rated lower than average (see the user rating below).

User Comment: "Just started in business; specializes in directories, pamphlets, paperback books or magazines; spiral binding or stitched; any size."

RATINGS	1	2	3	4	5	6	7	8	9	10	Ave	
Speed	-	-	1	-	-	-	-	-	-	-	---	1
Price	-	-	-	-	-	-	-	-	1		---	
Dependability	-	-	1	-	-	-	-	-	-	-	---	
Service . . .	-	-	1	-	-	-	-	-	-	-	---	
Quality . . .	-	-	-	1	-	-	-	-	-	-	---	
Overall . . .	-	-	-	-	1	-	-	-	-	-	---	

```
* * * * * * * * * * * * * * * * * * * * * * * * * * * * * * * *
*  Tip:  You might try to arrange a long-term contract with    *
*        one of your printers to produce a series of books for  *
*        you.  Ask them to quote on the entire job at once.     *
*        They should be able to give you a lower price because  *
*        such a long-term contract will allow them to make      *
*        fuller use of their facilities, save on quantity       *
*        purchases, and secure a more reliable cash flow.       *
* * * * * * * * * * * * * * * * * * * * * * * * * * * * * * * *
```

PRINTERS NOW OUT OF BUSINESS

The following printers are apparently out of business. If you are aware that any of them are still in business somewhere, please let us know. Thank you for your help.

Alpha Printing
6301 Central Avenue NW
Albuquerque NM 87105

AMP Publishing
111 N 10th St #B2
Olean NY 14760

Automation Book Printing
P O Box 12201
El Cajon CA 92022

Colonial Press
25 Broad St or 67 River St
Hudson MA 01749

Colwell
274 Fillmore Ave, Box 70037
St Paul MN 55107-0037
(taken over by Viking Press)

Daring Press
Dennis W. Bartow, General Manager
2020 Ninth Street SW
Canton OH 44706
(Daring is no longer printing books but is still publishing.

Open Studio
187 E Market Street
Rhinebeck NY 12572

H. Paul Publishing Company
Harry I. Paul, President
4883 Ronson Court, Suite L
San Diego CA 92111

Universal Lithographers
10626 York Road
Corkeyville MD 21030

Rumford National Graphics
475 Park Ave S
New York NY 10016
        or
71 S Central Ave
Valley Stream NY 11580
(516) 561-6300
        or
1 Bay Point
Meredith NH 03253
        or
Concord NH 03301

We've spent many hours trying to track this company down. Their New York and Meredith NH phone numbers are no longer in service. We do not have a street address for their Concord NH address, but there is no phone listing in Concord under Rumford. We called their Valley Stream NY phone number during ordinary business hours but received no answer. Yet, they are listed in the current issue of Direct Marketing magazine as a publications and catalog printer. They have never answered any of our RFQ's or printer surveys.

BOOK PRINTERS ON THE MOVE

The following book printers have moved from the addresses noted below. Our letters to them were returned by the post office as undeliverable and not forwardable. They may be out of business. Again, if you know of the status of any of these printers, please let us know.

Brooks Graphics
Jack Ferro, President
2901 Simms Street
Hollywood FL 33022

California Syllabus
1494 MacArthur Boulevard
Oakland CA 94602

Commercial Printing
221 N College
Fort Collins CO 80524

Fountain Press
3255 Quivas
Denver CO 80211

Graphicopy
P O Box 285
Floral Park NY 11001

New Publishing
P O Box 17068
Pittsburg PA 15235

Oberlin Printing Company
13500 W Lake Road
Vermilion OH 44089

Paper Faces
968 Third Avenue
New York NY 10022

Personalized Press
770 Spring Street NW
Atlanta GA 30308

Profile Press
Arlene McDonald, President
229 W 28th Street, 12th Floor
New York NY 10001

Terrace Press
P O Box 47
Bowmansville NY 14026

Triton Press
13850 Big Basin Way
Boulder Creek CA 95006
        or
325 Surf
Morro Bay CA 93442

MAIL ORDER BOOKLET PRINTERS

The following commercial printers can produce saddlestiched booklets (from 8 to 52 pages). Most of them advertise nationally and sell their printing services by mail; hence, they are used to working with customers via the U.S. postal service and UPS.

Of course, this is only a small sampling of those mail order printers who can produce saddlestitched booklets. We include only those whose work we have seen or whose price lists we have.

In many cases you may be able to obtain a better printing job locally (that means: a faster, less expensive, and better quality job). But if you live as we do among the cornfields of Iowa, mail order printers can save you money while doing the job just as fast and, in many cases, better than a local printer.

Send for the standard price lists of some of these printers. Then compare their prices with those you have obtained from your local printer. If the differences are small, you are better off using your local printer; but if the prices are significantly different, test a mail order printer by giving them a small order. Then, if they produce quality work and meet their deadlines, you might try them again.

Amity Hallmark
40-09 149th Place
P O Box 929, Linden Hall Stat.
Flushing NY 11354
(212) 939-2323

Champion Printing
Mike Stewart, Sales Manager
1677 Central Pkwy / P O 14129
Cincinnati OH 45214
(800) 543-1957
(513) 241-5233

J & G Printing
3511 Clark Road
Sarasota FL 33581
(813) 921-5508
"Good local printer with competitive advertising on booklet prices. Excellent work."

Nationwide Printing Inc.
518 Enterprise Dr / P O 17630
Fort Mitchell KY 41017
(606) 341-6446
"Fair work, 2 - 3 week turnaround time."

PAK Discount Printing
K. A. Porter, President
38771 N Lewis Ave
Zion IL 60099
(312) 249-1789
"Acceptable work, but slow."

Press America
Attn: New Order Department
5519 W Montrose Avenue
Chicago IL 60641
(312) 736-6569
"Good work, fast turnaround
(3 - 5 days), dependable."

Speedy Printers
Charles Margolis, President
23860 Miles Road
Cleveland OH 44128-5429
(216) 662-4141

HOW TO DEAL WITH OVERSEAS PRINTERS

We do not recommend having overseas printers (from Asia or Europe) do the printing for your books unless you have had years of experience in specifying your print requirements and in dealing with printers. Nevertheless, for those of you who are considering working with such printers, we will list some of the advantages and disadvantages of having your printing done overseas, and then give some guidelines on how to work most effectively with such printers.

The **advantages** of working with overseas printers are:

1) Overseas printers are usually cheaper, especially for full-color work and illustrated books.
2) They offer good quality, often superb full-color reproduction.
3) Hong Kong or Singapore printers can often give good prices for intricate typesetting,
4) Asian printers like long-term relationships and do whatever they can to develop satisfied customers.

The **disadvantages** are:

1) The manufacturer's clause of the copyright law says that you can import no more than 2000 copies of a any printed materials copyrighted by a U.S. citizen but manufactured outside the country. There are two ways you can get around this limit: If any part of the production is done in the U.S. (e.g., typesetting or binding), then the clause does not apply. It also does not apply to books of a "primarily non-textual nature" (e.g., photography books, illustrated books, travel guides). Check with the U.S. Customs Service for details regarding this law before proceeding.
2) Many of the higher quality foreign printers don't do short runs.
3) Frustrations can arise due to misunderstandings --
   * differences in specifications
   * language or terminology differences
   * quality of materials (especially paper).
4) Shipping time adds to usual time for production.
5) Communication can be more difficult --
   * higher cost of phone calls or telexes
   * time zone gap.
6) Piracy of copyrighted material can be a problem in Taiwan, Korea, the Philippines, and Indonesia because they do not abide by international copyright standards.
7) The high humidity of Southeast Asia limits the amount of paper they can keep in storage; hence, they tend to buy paper by the job (which can delay your job, especially if you require special paper).

8) The high humidity can also cause warping of perfectbound books, especially hardcovers, unless sewn first.
9) Many Asian printers favor long-term relationships with their customers; if you're new or only doing a one-time job with them, your job may be bumped for one of their steady customers.
10) Proofreading of typesetting done by foreign printers can delay your book by many weeks since it is not safe to leave it to the typesetters (where English is often a second language at best).

Guidelines -- Here are some tips on how to deal most effectively with overseas printers:

1) Clearly define all your specifications in writing.
2) Make sure the printer understands exactly what you expect.
3) Examine bids carefully for any differences from your specifications.
4) Check to see that the standard operating assumptions of the printing trade in that country match your own assumptions.
5) Get references. And check them out thoroughly.
6) Have them send you samples of their work. Inspect those samples very carefully.
7) Monitor all stages of the manufacturing process.
8) Be sure to get updated regularly.
9) When possible, work with their sales representatives in this country. Let them deal with the language barrier and time differences.

OVERSEAS PRINTERS — A LISTING

Please note that what follows is not a complete listing of all overseas printers. First, it would be impossible to make a complete list. Second, since we do not encourage publishers with short-run needs to use overseas printers, we have not made an exhaustive effort to research this field.

Overseas printers with U.S. sales offices or brokers:

Creative Press & Manufacturing
Hwan Kwak, General Manager
437 Madison Avenue
New York NY 10022
(Taiwan)

Dai Nippon
Ms. Regan Connolly, Sales Rep
1633 Broadway, 15th Floor
New York NY 10019
(Japan, Hong Kong)

David Haworth Associates
David Haworth, President
2550 M St NW #525
Washington DC 20037
(Finland)

Mandarin Offset Inc.
Cynthia M. Parzych, President
1501 Third Avenue
New York NY 10028
(Hong Kong, Singapore, China)

Mondadori-Ame Publishing Ltd.
Bruno Nicolis, Sales Manager
437 Madison Avenue
New York NY 10022
(Italy)

Nisha Printing
149 Madison Avenue
New York NY 10016
(Japan)

On Line Press
Jim Hwong, General Manager
3720 D Campus Drive
Newport Beach CA 92660
(Korea)

66 Litho USA
286 Main Avenue
Passaic NJ 07055

Toppan Printing Company
Mitsuhiro Tada, General Mgr.
680 Fifth Avenue
New York NY 10019
(Japan, Hong Kong)

Hong Kong Printers

Amsel Limited
130 Connaught Rd
Alliance Building Room 1602
Central, Hong Kong

Colorcraft
Daniel Chung - Director
18 Whitfield Road
Citicorp Centre - Room 502-3
Causeway Bay, Hong Kong

Dah Hua Printing Press Company
Bernard Chan - Sales Director
26 Lee Chung Street
Chaiwan Industrial Bldg 9/F
Chaiwan, Hong Kong

Emphasis Hong Kong Limited
Wilson House 10th Fl
19-27 Wyndham Street
Central, Hong Kong

Hong Kong Printers continued

Great Wall Printing & Graphics
Philip Rosenberg or Eliza Nip
30 Hollywood Road 11th Floor
Central, Hong Kong

Liang Yu Printing Factory Ltd.
Eric Hui, Deputy Mng Director
9-11 Sai Wan Ho St
Hip Shing Industrial Building
Shaukiwan, Hong Kong

Libra Press Limited
R Pennels - Director
56 Wong Chuk Hang Road 5D
Hong Kong

Linkprint Limited
Alfred Chan--Managing Director
51-53 Johnston Road
Shiu Fung Commercial Bldg 8/F
Wanchai, Hong Kong

Pacific Offset Printing Co Ltd
8 Dockyard Lane
Block A / Wah Ha Factory 6/F
Quarry Bay, Hong Kong

South Sea International Press
Unit 3 - 20th Floor
Eastern Centre Bldg.
1065 King's Road
Quarry Bay, Hong Kong

Travel Publishing Asia Ltd.
Roy Howard, Managing Director
1801 World Trade Centre
Causeway Bay, Hong Kong

Wing King Tong Company Limited
41-55 Wo Tong Tsui
Wing Foo Industrial Bldg 3/F
Kwai Chung N.T., Hong Kong

Yee Tin Tong Printing Press
Jim Viney
Tong Chong St 4th FL
South China Morning Post Bldg.
Quarry Bay, Hong Kong

Singapore Printers

International Press Company
Tan Kim Hui, Export Mktg Mgr.
32 Kallang Place
Singapore 1233, Singapore

Kok Wah Press Pte Ltd.
3 Gul Crescent
Singapore 2262, Singapore

Koon Wah Printing Pte Ltd
18 Tuas Avenue 5
Singapore 2263, Singapore

Pac Press Industries Pte Ltd.
25 Tannery Lane #01-25/27
Singapore 1334, Singapore

Singapore National Printers
303 Upper Serangoon Road
Singapore 1334, Singapore

Tien Wah Printing
Sally Pang - Sales Manager
977 Bukit Timah Road
Singapore 2158, Singapore

Taiwan Printers

Printechnic Associates Inc.
P O Box 11011
Taipei, Taiwan

Taipei Yung Chang Printing Co.
49-15 Chuanyuan Road
Peitou (112)
Taipei, Taiwan

For more information regarding printers in Asia, write to the following places:

Guides

1) A Publisher's Guide to Printing in Asia, 1985 Edition - $8.00

   Travel Publishing Asia Ltd
   1801 World Trade Centre
   Causeway Bay, Hong Kong

   Note: Most of the above information on Asian printers came
         from this excellent guide. The $8.00 covers air mail.

2) Printing in Asia - $5.00

   Moon Publications
   P O Box 1696
   Chico CA 95927

   Note: We haven't seen this guide, but it has been
          recommended to us by others.

Lists

You can obtain free lists of printers in various countries by writing to their board of trade. Here are some addresses you can write to for more information:

HONG KONG

Hong Kong Development Council
Great Eagle Centre, 31st Floor
Harbour Road, Hong Kong
Phone: 5-8334333
New York: 212-730-7777

Hong Kong Printers Association
48 Johnston Road, 1st Floor
Wanchai, Hong Kong
Phone: 5-275050

JAPAN

Japan Book Publishers Assn.
6 Fukuro-machi, Shinjuku-ku
Tokyo 162, Japan
Phone: (03) 2681301

Japan Trade Center
New York NY
Phone: 212-997-0432

SINGAPORE

Master Printers Association
04-02 Association Building
68 Lorong 16 Geylang
Singapore 1439, Singapore
Phone: 745619

Singapore Trade Development Bd
350 S Figueroa St
Los Angeles CA 90071
Phone: 212-617-7358

CLASSIFIED DIRECTORY OF SHORT-RUN BOOK PRINTERS

The following classified lists are derived from the printer survey forms which we sent out in late spring of 1985. Only those printers who actually completed the survey forms and returned them to us are listed. We hope these lists will make it easier for you to find those printers best prepared to meet your book printing needs.

Ultra-Short-Run Specialists — under 1000 copies (optimum)

Academy Books
C & M Press
Coach House Press
Coneco Laser Graphics
Crane Duplicating Service Inc.
Giant Horse and Company
GRT Book Printing
Independent Printing Company
The Job Shop

Morgan Printing
National Reproductions Corp.
Omnipress
Quinn-Woodbine
Readi Multi-Lith
Repro-Tech
Van Volumes
Westview Press

Short-Run Specialists — 1000 to 5000 copies (optimum)

Banta Company
Best Impressions
T. H. Best Printing Company
Blake Printery
Book-Mart Press
Bookcrafters
Braun-Brumfield Inc.
Caldwell Printers
Canterbury Press
Capital City Press
Comput-A-Print
Corley Printing Company
Cushing-Malloy
Delta Lithograph
R. R. Donnelley & Sons Company
Eastern Lithographing
Edwards Brothers
Faculty Press
Fay Printing Center
Germac Printing
Gilliland's Printing Company
Halliday Lithograph
Henington Publishing Company
Hignell Printing Ltd.
Inter-Collegiate Press

Kansas City Press
Kimberly Press
Kni Book Manufacturing
Malloy Lithographing
McGregor & Werner
McNaughton & Gunn
Meriden-Stinehour
Metromail
Mitchell-Shear
Nimrod Press
Paraclete Press
Patterson Printing Company
Paust Incorporated
The Print Center
Publishers Press
Schiff Printers & Lithographer
Service Printing Company
Sexton Printing
John S. Swift Company
Thomson-Shore Inc.
Town House Press
Unicorn Press
Walsworth Publishing
Whitehall Printing
Wickersham Printing Company

Printers Specializing in Serving Self-Publishing Authors

Adams Press
Andover Press
Apollo Books
Dynamic Printing
Exposition Press of Florida
Geryon Press
Harlo Printing Company
Heart of the Lakes Publishing

Lorrah & Hitchcock Publishers
Maverick Publications
McClain Printing Company
Morgan Printing
Prinit Press
Snohomish Publishing
Southern Tennessee Publishing
Torch Publications

Typesetters -- Printers with typesetting capabilities

* Astericks indicate those with telecommunications/typesetting.

Academy Books
* Access Composition Services
American Lithocraft Corp.
Apollo Books
* Automated Graphic Systems
* Bawden Printing
Bay Port Press
* Ovid Bell Press
Best Impressions
Blake Printery
Book-Mart Press
* Bookcrafters
William Boyd Printing
Braun-Brumfield Inc.
Brennan Printing
* R. L. Bryan Company
Caldwell Printers
Canterbury Press
* Capital City Press
* Central Publishing Company
* Coach House Press
* Colortone Press
* Commercial Printing Company
Community Press
Comput-A-Print
* Coneco Laser Graphics
Contemporary Lithographers
* Crane Duplicating Service
* Delta Lithograph
R. R. Donnelley & Sons
Edwards Brothers
Evangel Press
Faculty Press
* Fay Printing Center
* Fort Orange Press
Friesen Printers

Futura Printing
Ganis & Harris
* George Lithograph
Germac Printing
Giant Horse and Company
* Gilliland's Printing Company
Golden Horn Press
* Harlo Printing Company
Hignell Printing Ltd.
* A. B. Hirschfeld Press
Hooven-Dayton Corporation
Independent Printing Company
Interstate Book Manufacturers
Interstate Printers and Publrs
* The Job Shop
* Jostens Printing & Publishing
Kingsport Press
Kni Book Manufacturing
* W. A. Krueger
Liberty York Graphic Industrie
John D. Lucas Printing Company
* Mack Printing
Maple-Vail Book Manufacturing
* Maverick Publications
* McClain Printing Company
* Meaker the Printer
* Meriden-Stinehour
Metromail
Mitchell-Shear
Morgan Printing
Morningrise Printing
* Neibauer Press
* Nimrod Press
O'Neil Data Systems
Oaks Printing Company
Paraclete Press

Typesetting Printers continued

Paust Incorporated
Pennysaver Press Inc.
Plain Talk Publishing Company
Port City Press
* Premier Printing Corporation
Prinit Press
The Print Center
Printing Corp. of America
Publishers Choice Book Mfg.
* Readi Multi-Lith
* Recorder Sunset Press
* Rose Printing Company
* Schiff Printers & Lithographer
Scribner Graphic Press
* Sexton Printing
* The Sheridan Press
Skillful Means Press
* Snohomish Publishing
Southeastern Printing Company

Southern Tennessee Publishing
* The Studley Press
John S. Swift Company
Torch Publications
Town House Press
* Tracor Publications
TSO General Corporation
* Unicorn Press
Victor Graphics
* Vogue Printers
* Walsworth Publishing
Walter's Publishing
West Side Graphics
* Westview Press
* White Arts Inc.
Wickersham Printing Company
Wood & Jones
Worzalla Publishing Company

Hardcovers -- Printers with in-house casebinding capability

Academy Books
Apollo Books
Banta Company
T. H. Best Printing Company
Bookcrafters
Braun-Brumfield Inc.
C & M Press
Capital City Press
Commercial Printing Company
Community Press
Contemporary Lithographers
R. R. Donnelley & Sons Company
Edwards Brothers
Eerdmans Printing Company
Faculty Press
Fairfield Graphics
Fort Orange Press
Ganis & Harris
Germac Printing
Hignell Printing Ltd.

Inter-Collegiate Press
Interstate Book Manufacturers
Interstate Printers and Publrs
Jostens Printing & Publishing
Kimberley Press
Kingsport Press
C. J. Krehbiel Company
W. A. Krueger
Maple-Vail Book Manufacturing
Meriden-Stinehour
Murray Printing Company
Optic Graphics Inc.
Publishers Press
Quinn-Woodbine
Recorder Sunset Press
Rose Printing Company
Skillful Means Press
Walsworth Publishing
Worzalla Publishing Company

4-Color Specialists — Short-Runs

Academy Books
Access Composition Services
Accurate Web
Adams Press
American Lithocraft Corp.
Apollo Books
Banta Company
Bawden Printing
Bay Port Press
Best Impressions
T. H. Best Printing Company
Blake Printery
Bookcrafters
William Boyd Printing Company
R. L. Bryan Company
Canterbury Press
CBP Press
Central Publishing Company
Coach House Press
Colortone Press
Commercial Printing Company
Community Press
Contemporary Lithographers
Crest Litho
Delta Lithograph
Dickinson Press
R. R. Donnelley & Sons Company
Eastern Lithographing
Eerdmans Printing Company
Evangel Press
Faculty Press
Fay Printing Center
Fort Orange Press
Friesen Printers
Futura Printing
Ganis & Harris
General Offset Company
Giant Horse and Company
Gilliland's Printing Company
Harlo Printing Company
Hignell Printing Ltd.
A. B. Hirschfeld Press
Inter-Collegiate Press
Interstate Printers and Publrs
Jostens Printing & Publishing
Julin Printing
Kimberly Press
Kingsport Press
C. J. Krehbiel Company
W. A. Krueger
Lithocolor Press

Little River Press
John D. Lucas Printing Company
Mack Printing
Maverick Publications
McClain Printing Company
Meaker the Printer
Meriden-Stinehour
Mitchell-Shear
Morgan Printing
Neibauer Press
Nimrod Press
Oaks Printing Company
Paraclete Press
Patterson Printing Company
Paust Incorporated
Phillips Brothers Printing
Plain Talk Publishing Company
Port City Press
Premier Printing Corporation
Prinit Press
The Print Center
Printing Corp. of America
Publishers Choice Book Mfg.
Publishers Press
Recorder Sunset Press
Rose Printing Company
Schiff Printers & Lithographer
Scribner Graphic Press
Semline Inc.
Service Printing Company
Sexton Printing
The Sheridan Press
Skillful Means Press
Snohomish Publishing
Southeastern Printing Company
Southern Tennessee Publishing
Spilman Printing Company
Straus Printing Company
The Studley Press
John S. Swift Company
Thomson-Shore Inc.
Torch Publications
Town House Press
Tracor Publications
Unicorn Press
Victor Graphics
Vogue Printers
Walsworth Publishing
White Arts Inc.
Wood & Jones
Worzalla Publishing Company

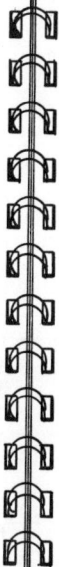

Catalog Printers — Short-Runs (one, two, or four-colors)

* Astericks indicate those printers who can also print corporate and association **annual reports**.

* Academy Books
* Access Composition Services
  Accurate Web
  Adams Press
* American Lithocraft Corp.
* Apollo Books
  Automated Graphic Systems
  Banta Company
* Bawden Printing
* Bay Port Press
* Ovid Bell Press
* Best Impressions
  T. H. Best Printing Company
* Blake Printery
  Book-Mart Press
  Bookcrafters
  William Boyd Printing Company
* Brennan Printing
* R. L. Bryan Company
* C & M Press
  Caldwell Printers
* Canterbury Press
  Capital City Press
* CBP Press
* Central Publishing Company
* Colortone Press
* Commercial Printing Company
* Community Press
* Comput-A-Print
* Coneco Laser Graphics
  Corley Printing Company
* Crane Duplicating Service
* Crest Litho
  Delta Lithograph
* Dickinson Press
* Dinner & Klein
  R. R. Donnelley & Sons Co.
* Dynamic Printing
  Eastern Lithographing
  Eerdmans Printing Company
* Faculty Press
* Fay Printing Center
* Fort Orange Press
  Friesen Printers
* Futura Printing
  George Lithograph
* Germac Printing

* Geryon Press
* Giant Horse and Company
  Gilliland's Printing Company
* Golden Horn Press
* Harlo Printing Company
* Hignell Printing Ltd.
* A. B. Hirschfeld Press
* Hooven-Dayton Corporation
* Independent Printing Company
  Inter-Collegiate Press
  Interstate Book Manufacturers
* Interstate Printers and Publrs
* The Job Shop
* Jostens Printing & Publishing
* Julin Printing
* Kingsport Press
  C. J. Krehbiel Company
* W. A. Krueger
* Letternation
* Liberty York Graphic
  Lithocolor Press
* Little River Press
* John D. Lucas Printing Company
* Mack Printing
  Maple-Vail Book Manufacturing
  Maverick Publications
* McClain Printing Company
  McNaughton & Gunn
  Meaker the Printer
* Meriden-Stinehour
* Mitchell-Shear
  Morgan Printing
  Morningrise Printing
  Murray Printing Company
  National Reproductions Corp.
* Neibauer Press
* Nimrod Press
  O'Neil Data Systems
* Oaks Printing Company
* Optic Graphics Inc.
* Paraclete Press
* Paust Incorporated
  Pennysaver Press Inc.
* Phillips Brothers Printing
* Plain Talk Publishing Company
* Premier Printing Corporation
  Prinit Press

Catalog and Annual Report Printers — Short Runs continued

* The Print Center
* Printing Corp. of America
* Publishers Choice Book Mfg.
* Publishers Press
* Quintessence Press
  Readi Multi-Lith
* Rose Printing Company
* Schiff Printers & Lithographer
* Scribner Graphic Press
  Semline Inc.
  Service Printing Company
* Sexton Printing
  The Sheridan Press
* Skillful Means Press
  Snohomish Publishing
* Southeastern Printing Company
* Spilman Printing Company
* Straus Printing Company
* The Studley Press

* John S. Swift Company
  Thomson-Shore Inc.
* Torch Publications
  Town House Press
* Tracor Publications
* TSO General Corporation
* Unicorn Press
  Versa Press
* Victor Graphics
* Vogue Printers
  Walsworth Publishing
* West Side Graphics
* Westview Press
* White Arts Inc.
  Whitehall Printing
  Wickersham Printing Company
* Wood & Jones
  Worzalla Publishing Company

Cookbook Printers — Short-Runs

The following four printers specialize in printing cookbooks for fundraisers. They offer standard formats and special help.

Brennan Printing
Cookbook Publishers, a division of Kansas City Press
Fundcraft
Walter's Cookbooks

Academy Books
Access Composition Services
Accurate Web
Adams Press
American Lithocraft Corp.
Andover Press
Apollo Books
Automated Graphic Systems
Bawden Printing
Best Impressions
T. H. Best Printing Company
Blake Printery
Bookcrafters
William Boyd Printing Company
Brennan Printing
Canterbury Press
Capital City Press
CBP Press
Central Publishing Company

Colortone Press
Community Press
Coneco Laser Graphics
Crane Duplicating Service Inc.
Delta Lithograph
Dickinson Press
R. R. Donnelley & Sons Company
Eerdmans Printing Company
Evangel Press
Exposition Press of Florida
Faculty Press
Fay Printing Center
Fort Orange Press
Friesen Printers
Fundcraft
Futura Printing
Geryon Press
Harlo Printing Company
Hignell Printing Ltd.

Cookbook Printers — Short Runs continued

A. B. Hirschfeld Press
Hooven-Dayton Corporation
Inter-Collegiate Press
Interstate Printers and Publrs
Jostens Printing & Publishing
Julin Printing
Kansas City Press
Kingsport Press
Lorrah & Hitchcock Publishers
Maple-Vail Book Manufacturing
Maverick Publications
McGregor & Werner
McNaughton & Gunn
Meaker the Printer
Meriden-Stinehour
Mitchell-Shear
Morgan Printing
Morningrise Printing
National Reproductions Corp.
Neibauer Press
Nimrod Press
Oaks Printing Company
Phillips Brothers Printing
Prinit Press

Printing Corp. of America
Publishers Press
Quintessence Press
Rose Printing Company
Scribner Graphic Press
Semline Inc.
The Sheridan Press
Skillful Means Press
Snohomish Publishing
Southeastern Printing Company
Thomson-Shore Inc.
Torch Publications
Town House Press
Tracor Publications
Van Volumes
Versa Press
Victor Graphics
Vogue Printers
Walsworth Publishing
Walter's Publishing
Westview Press
Wickersham Printing Company
Worzalla Publishing Company

Demand Printers -- Allows for smaller quantities to be reprinted without paying a premium rate -- the reruns are charged at the same per copy rate as the initial order (and can be done almost right away). Check with the printers listed below to verify their services.

Access Composition Services
Best Impressions
C & M Press
Comput-A-Print
Coneco Laser Graphics
Data Copi
Fay Printing Center
Geryon Press
Independent Printing Company
The Job Shop
Lettermation

McGregor & Werner
Meriden-Stinehour
National Reproductions Corp.
Oaks Printing Company
Plain Talk Publishing Company
Quintessence Press
Readi Multi-Lith
Southeastern Printing Company
John S. Swift Company
Westview Press
White Arts Inc.

Galleys -- Printers capable of producing reader's galley copies.

Access Composition Services
Apollo Books
Best Impressions
Central Publishing Company
Comput-A-Print
Coneco Laser Graphics
Crane Duplicating Service Inc.
Fort Orange Press
Geryon Press
Independent Printing Company
Kimberly Press
McClain Printing Company
McGregor & Werner
Meriden-Stinehour
Mitchell-Shear

National Reproductions Corp.
Oaks Printing Company
Paraclete Press
Paust Incorporated
Plain Talk Publishing Company
The Print Center
Quintessence Press
Schiff Printers & Lithographer
Southern Tennessee Publishing
Tracor Publications
Van Volumes
Walsworth Publishing
Westview Press
Wood & Jones

Journal Printers (and Magazines) — Short Runs

* Astericks indicate those printers who can also print magazines in one, two, or four colors.

* Academy Books
* Access Composition Services
* American Lithocraft Corp.
* Apollo Books
  Automated Graphic Systems
* Bawden Printing
* Bay Port Press
* Ovid Bell Press
* Best Impressions
* Blake Printery
  Book-Mart Press
  Bookcrafters
* William Boyd Printing Company
  Braun-Brumfield Inc.
* R. L. Bryan Company
* C & M Press
* Canterbury Press
* Capital City Press
* CBP Press
* Central Publishing Company
  Coach House Press
  Commercial Printing Company
* Community Press
  Comput-A-Print
  Coneco Laser Graphics
  Crane Duplicating Service
* Crest Litho
  Cushing-Malloy

* Delta Lithograph
* R. R. Donnelley & Sons Company
* Dynamic Printing
  Eastern Lithographing
  Edwards Brothers
  Evangel Press
* Faculty Press
* Fay Printing Center
* Fort Orange Press
* Futura Printing
  Ganis & Harris
  George Lithograph
  Giant Horse and Company
  Gilliland's Printing Company
* Harlo Printing Company
  Heart of the Lakes Publishing
* Hignell Printing Ltd.
* A. B. Hirschfeld Press
* Hooven-Dayton Corporation
* The Sheridan Press
* Skillful Means Press
* Southeastern Printing Company
  Southern Tennessee Publishing
  Spilman Printing Company
* Straus Printing Company
* The Studley Press
  Thomson-Shore Inc.
  Town House Press

* Tracor Publications
  TSO General Corporation
  Independent Printing Company
* Inter-Collegiate Press
* Interstate Printers and Publrs
* Jostens Printing & Publishing
* Kimberly Press
* Kingsport Press
  Kni Book Manufacturing
* W. A. Krueger
  Liberty York Graphic
* Lithocolor Press
* Little River Press
* John D. Lucas Printing Company
* Mack Printing
  Malloy Lithographing
  Maverick Publications
* McClain Printing Company
  McGregor & Werner
  McNaughton & Gunn
* Meaker the Printer
* Meriden-Stinehour
* Mitchell-Shear
* Morgan Printing
  National Reproductions Corp.
* Neibauer Press
* Nimrod Press
* Oaks Printing Company
  Omnipress

* Paraclete Press
* Patterson Printing Company
* Paust Incorporated
* Pennysaver Press
  Phillips Brothers Printing
* Plain Talk Publishing Company
* Premier Printing Corporation
* Prinit Press
* The Print Center
* Printing Corp. of America
* Publishers Choice Book Mfg.
* Publishers Press
  Quinn-Woodbine
  Quintessence Press
  Readi Multi-Lith
* Rose Printing Company
* Schiff Printers & Lithographer
* Scribner Graphic Press
* Sexton Printing
* Unicorn Press
  Van Volumes
  Victor Graphics
* Walsworth Publishing
* West Side Graphics
* Westview Press
  White Arts Inc.
  Wickersham Printing Company
  Wood & Jones
* Worzalla Publishing Company

Mass-Market -- Printers capable of printing books about the
size of mass-market paperbacks (4 1/8" x 7")

American Lithocraft Corp.
Apollo Books
Banta Company
Bookcrafters
Braun-Brumfield Inc.
Crest Litho
Delta Lithograph
Eastern Lithographing
Edwards Brothers
Eerdmans Printing Company
Ganis & Harris
Giant Horse and Company
Golden Horn Press
Harlo Printing Company
Inter-Collegiate Press
Interstate Book Manufacturers

Jostens Printing & Publishing
Kingsport Press
Kni Book Manufacturing
Lithocolor Press
Little River Press
McNaughton & Gunn
Offset Paperback Manufacturers
Paust Incorporated
Prinit Press
Printing Corp. of America
Publishers Press
Rose Printing Company
Town House Press
Van Volumes
Versa Press
Wickersham Printing Company

Picture Books -- Printers able to produce color illustrated books for children (either hardcover or softcover)

Access Composition Services
Accurate Web
American Lithocraft Corp.
Apollo Books
Best Impressions
T. H. Best Printing Company
Bookcrafters
William Boyd Printing Company
Central Publishing Company
Colortone Press
Community Press
Delta Lithograph
Dickinson Press
R. R. Donnelley & Sons Company
Eastern Lithographing
Faculty Press
Fort Orange Press
Friesen Printers
Futura Printing
General Offset Company
Harlo Printing Company

Hignell Printing Ltd.
Jostens Printing & Publishing
Kingsport Press
C. J. Krehbiel Company
W. A. Krueger
Little River Press
Meriden-Stinehour
Mitchell-Shear
Nimrod Press
Oaks Printing Company
Paraclete Press
Phillips Brothers Printing
Printing Corp. of America
Rose Printing Company
Semline Inc.
Skillful Means Press
Southeastern Printing Company
Tracor Publications
Victor Graphics
Walsworth Publishing
Worzalla Publishing Company

Yearbook Printers -- Printers with the capability to produce fancy school and military yearbooks.

Note: Hunter, Jostens, and Walsworth have divisions specializing in producing yearbooks for high schools and colleges.

T. H. Best Printing Company
William Boyd Printing Company
Capital City Press
Community Press
Eastern Lithographing
Friesen Printers
Hignell Printing Ltd.

Hunter Publishing Company
Inter-Collegiate Press
Jostens Printing & Publishing
Kingsport Press
Semline Inc.
Town House Press
Walsworth Publishing

BOOK PRINTERS SORTED ALPHABETICALLY BY STATE AND CITY

AL	Birmingham	Oxmoor Press	205-942-0511
AZ	Phoenix	Access Composition Services	602-272-7778
	Phoenix	Meaker the Printer	602-254-2171
	Phoenix	Messenger Graphics	602-254-7231
	Scottsdale	W. A. Krueger	602-948-5650
	Tucson	Dynamic Printing	602-883-5610
	Tucson	Fabe Litho Ltd.	602-622-2857
CA	Amador City	Quintessence Press	209-267-5470
	Anaheim	Kni Book Manufacturing	714-956-7300
	Arcadia	Caldwell Printers	818-447-4601
	Berkeley	Consolidated Printers	415-843-8524
	Berkeley	Golden Horn Press	415-845-4355
	Berkeley	West Coast Print Center	415-849-2746
	Brisbane	Foote & Davies	415-467-7100
	Campbell	Letternation	408-559-6577
	Costa Mesa	Morningrise Printing	714-957-8494
	Daly City	Giant Horse and Company	415-468-0573
	Emeryville	Albany Press	415-428-1800
	Fullerton	Premier Printing Corporation	714-871-3121
	Glendale	Griffin Printing & Lithograph	818-244-2128
	Goleta	Kimberly Press	805-964-6469
	Happy Camp	Naturegraph Publishers Inc.	916-493-5353
	Hollywood	Hal Leighton Printing Company	213-983-1105
	Los Angeles	American Offset Printers	213-231-4133
	Los Angeles	O'Neil Data Systems	213-820-4247
	Mountain View	Scribner Graphic Press	415-967-8118
	National City	Bay Port Press	619-420-6296
	Nevada City	Harold Berliner	916-273-2278
	Nevada City	Pelican Pond Publishing	?
	Oakland	GRT Book Printing	415-534-5032
	Oakland	Skillful Means Press	415-839-3931
	Pasadena	Wood & Jones	213-681-9663
	Sacramento	Spilman Printing Company	916-448-3511
	San Diego	Torch Publications	619-299-2111
	San Francisco	Economy Bookcraft	415-362-2708
	San Francisco	George Lithograph	415-397-2400
	San Francisco	Recorder Sunset Press	415-621-5400
	San Francisco	Synthex Press	415-824-8282
	San Leandro	Service Printing Company	415-352-7890
	San Luis Obispo	Blake Printery	805-543-6843
	Santa Ana	Unicorn Press	714-546-7320
	Van Nuys	Delta Lithograph	213-873-4910
	Visalia	Jostens -- Visalia	209-651-3300
CO	Boulder	Johnson Publishing Company	303-443-1576
	Boulder	Westview Press	303-444-3541
	Denver	Access Composition -- Denver	303-458-6955
	Denver	Eastwood Printing & Publishing	303-296-1905
	Denver	Golden Bell Press	303-296-1600
	Denver	A. B. Hirschfield Press	303-320-8500
	Denver	Quality Press	303-761-2160

CO	Denver	Readi Multi-Lith	303-987-2338
	Thornton	C & M Press	303-289-4757
CT	Bloomfield	Connecticut Printers	203-242-0711
	Meriden	Meriden-Stinehour	203-235-7929
	New Haven	Eastern Press	203-777-2353
	Norwalk	Arcata Graphics	800-722-7020
	Old Saybrook	The Saybrook Press	203-388-5737
	Stamford	Ray Freiman and Company	203-322-2474
	Weston	Eastern Publishing Graphics	?
DC	Washington	Colortone Press	202-387-6800
	Washington	Lanman Company	202-269-5400
	Washington	McGregor & Werner	202-722-2200
	Washington	Offset Composition Services	202-783-1010
FL	Boynton Beach	Futura Printing	305-734-0825
	Miami	Little River Press	305-757-7504
	Orlando	Rollins Press Inc.	305-677-5533
	Pompano Beach	Exposition Press of Florida	305-979-3200
	Pompano Beach	Printing Corp. of America	305-781-8100
	Stuart	Southeastern Printing Company	305-287-2141
	Tallahassee	Rose Printing Company	904-576-4151
GA	Atlanta	W. R. Bean & Son	404-691-5020
	Atlanta	Data Copi	404-261-0133
IA	Deep River	Brennan Printing	515-595-2000
	Des Moines	Plain Talk Publishing Company	515-282-0483
	Eldridge	Bawden Printing	319-285-4800
	Monticello	Julin Printing	319-465-3558
	Newell	Bireline Publishing Company	712-272-4417
IL	Bloomington	Pantagraph Printing	309-829-1071
	Chicago	Adams Press	312-676-3426
	Chicago	AOS Publishing Services	312-782-6722
	Chicago	W. F. Hall Inc.	312-794-4600
	Chicago	Micro Book Manufacturing Co.	312-922-2083
	Chicago	Rand McNally & Company	312-267-6868
	Chicago	Regensteiner Press	312-666-4200
	Chicago	Standard Printing Service	312-346-0499
	Chicago	John S. Swift -- Chicago	312-666-7070
	Danville	Interstate Printers and Publrs	217-446-0500
	East Dubuque	Johnson Graphics	815-747-6511
	East Peoria	Versa Press	309-822-8272
	Hillside	Wallace Press	312-626-2000
	Milan	Desaulniers Printing Company	309-799-7331
	North Chicago	Vogue Printers	312-689-4044
	Park Ridge	Franklin Press	312-648-1512
	Skokie	Great Northern Design	312-674-4740
	Springfield	Phillips Brothers Printing	800-637-9444
	Westchester	Lithocolor Press	312-345-5530
	Wheeling	Whitehall Printing	312-541-9290
	Woodstock	D. B. Hess Company	815-338-6900
IN	Crawfordsville	R. R. Donnelley & Sons Company	800-428-0832
	Dublin	Prinit Press	317-478-4885
	Indianapolis	Central Publishing Company	317-636-4504
	Indianapolis	White Arts Inc.	317-638-3564

IN	Middlebury	Yoder's Printing	no phone
	Nappanee	Evangel Press	219-773-3164
	Richmond	Paust Incorporated	317-962-1507
	Speedway	Carl Hungness Publishing	317-244-4792
KS	Arkansas City	Gilliland's Printing Company	800-332-8200
	Olathe	Interstate Book Manufacturers	800-255-0003
	Olathe	Kansas City Press	800-821-5745
	Shawnee Mission	Inter-Collegiate Press	913-432-8100
	Topeka	Jostens -- Topeka	913-266-3300
	Wichita	Letters Inc.	?
KY	Louisville	Courier Graphics	502-458-5303
	Murray	Lorrah & Hitchcock Publishers	502-753-3759
	Utica	McDowell Publications	?
LA	Baton Rouge	Moran Colorgraphic	504-923-2550
MA	Acton	Concepts Unlimited	617-263-6777
	Avon	Lorell Press	617-471-7750
	Barnstable	Crane Duplicating Service Inc.	617-362-2700
	Boston	Adams & Abbott Inc.	617-542-1621
	Boston	Nimrod Press	617-437-7900
	Braintree	Semline Inc.	617-848-2380
	Dalton	The Studley Press	413-686-0441
	Lawrence	Graphic Litho Corporation	617-683-2766
	Lexington	Lexington Press	617-862-8900
	Middleborough	Country Press	617-947-4485
	Orleans	Paraclete Press	617-255-4685
	Randolph	Modern Graphics	617-986-4262
	Stoughton	Alpine Press	800-343-5901
	West Hanover	Halliday Lithograph	617-826-8385
	Westford	Murray Printing Company	617-692-6321
	Wilbraham	Van Volumes	413-596-2113
	Winchester	University Press	617-729-8000
	Woods Hole	The Job Shop	617-548-9600
	Worcester	Heffernan Press	617-791-3661
MD	Baltimore	Graphic Offset	301-539-8306
	Baltimore	John D. Lucas Printing Company	301-633-4200
	Baltimore	Port City Press	301-486-3000
	Baltimore	Victor Graphics	301-233-8300
	Baltimore	Waverly Press	800-638-0673
	Beltsville	Columbia Planograph	301-937-4677
	Easton	Economy Printing Company	?
	Glen Burnie	Optic Graphics Inc.	800-638-7107
	Rockville	Holladay Tyler Printing Corp.	301-881-8050
	White Plains	Automated Graphic Systems	301-843-1800
ME	Lewiston	Twin City Printery	207-784-9181
	Norway	Oxford Group	207-743-8953
	Portland	Anthoensen Press	207-774-3301
MI	Ann Arbor	Braun-Brumfield Inc.	313-662-3291
	Ann Arbor	Cushing-Malloy	313-663-8554
	Ann Arbor	Edwards Brothers	313-769-1000
	Ann Arbor	Malloy Lithographing	800-722-3231
	Ann Arbor	Mitchell-Shear	313-995-2505
	Benton Harbor	Patterson Printing Company	616-925-2177

MI	Berrien Springs	University Printers	616-471-3236
	Chelsea	Bookcrafters	313-475-9145
	Detroit	Harlo Printing Company	313-883-3600
	Detroit	National Reproductions Corp.	313-961-5252
	Dexter	Thomson-Shore Inc.	313-426-3939
	Grand Rapids	Dickinson Press	616-451-2957
	Grand Rapids	Eerdmans Printing Company	616-451-0763
	Grand Rapids	Four Corners Press	616-243-2015
	Lansing	John Henry Company	517-484-5403
	Saline	McNaughton & Gunn	313-429-5411
MN	Duluth	Davidson Printing Company	218-727-8721
	Eden Prairie	Viking Press	?
	Minneapolis	Bolger Publications	612-645-6311
	Minneapolis	Haymarket Press	612-721-4401
	Minneapolis	Jostens Printing & Publishing	612-830-8415
	St Paul	The Webb Company	612-690-7200
	Waseca	Walter's Publishing	800-447-3274
	Winona	Apollo Books	800-328-8963
MO	Fulton	Ovid Bell Press	800-835-8919
	Independence	Independence Press	816-252-5010
	Marceline	Walsworth Publishing	816-376-3543
	Ozark	William Yates / Printer	?
	Springfield	Fay Printing Center	417-883-1520
	St Louis	CBP Press	314-371-6900
	St Louis	Corley Printing Company	314-426-3900
	St Louis	Plus Communications	?
	St Louis	John S. Swift Company	314-991-4300
	St Louis	Universal Printing Company	314-771-6900
	St Louis	Von Hoffman Press	314-966-0909
MS	Senatobia	W. A. Krueger -- Senatobia	601-562-5252
NC	Charlotte	The Delmar Company	704-847-9801
	Charlotte	Heritage Printers	704-372-5784
	Lillington	Edwards Brothers -- Raleigh	919-893-2717
	Raleigh	Contemporary Lithographers	919-821-2211
	Reidsville	Omega Industry	?
	Winston-Salem	Hunter Publishing Company	?
NE	Lincoln	Metromail	402-475-4591
	Omaha	Interstate Printing Company	402-341-8028
NH	Berlin	Oxford Group -- New Hampshire	603-752-2339
	Grantham	Tompson & Rutter Inc.	603-863-4392
NJ	Bayonne	Jersey Printing Company	201-436-4200
	Carlstadt	Davis Printing Corporation	201-935-5100
	Cedar Grove	Rae Publishing Company	201-239-1600
	Englewood Cliff	Xerographic Reproduction Ctr.	201-871-4011
	Hoboken	Guinn Printing Company	201-659-9000
	Jersey City	American Lithocraft Corp.	201-434-6617
	Jersey City	General Offset Company	201-420-0500
	Jersey City	Mansfield Chart Service	201-795-0629
	North Bergen	Book-Mart Press	201-864-1887
	Phillipsburg	Harmony Press	?
	Rutherford	Multiprint Inc.	201-935-7474
	Summit	Yarrow Inc.	201-273-8390

NJ	Teterboro	John S. Swift -- Teterboro	201-288-2050
	West Patterson	Repro-Tech	201-785-0011
	Woodbine	Quinn-Woodbine	609-861-5352
NV	Reno	Comput-A-Print	702-786-2300
NY	Albany	William Boyd Printing Company	518-436-9686
	Albany	Crest Litho	518-456-2296
	Albany	Fort Orange Press	518-489-3233
	Bronx	PPI Press	212-292-5536
	Brooklyn	Copen Press	212-235-4270
	Brooklyn	Faculty Press	718-851-6666
	Brooklyn	The Print Center	212-206-8465
	Chester	Goshen Litho	914-469-2102
	College Point	Algen Press Corporation	212-463-4605
	Deer Park	Accurate Web	516-667-3200
	Deer Park	TGI Graphics	516-586-1973
	Dobbs Ferry	Morgan Press	914-693-0023
	Farmingdale	B. C. Graphics	516-293-9136
	Farmingdale	Coleman Graphics	516-293-0383
	Farmingdale	Northeast Web Printing	516-454-1600
	Glen Falls	Coneco Laser Graphics	518-793-3823
	Hempstead	Liberty York Graphic Industrie	516-481-8500
	Interlaken	Heart of the Lakes Publishing	607-532-4997
	Ithaca	Willcox Press	607-272-1212
	Jericho	Pennysaver Press Inc.	516-997-7755
	Long Island City	Peter F. Mallon Inc.	212-786-2000
	Long Island City	TSO General Corporation	718-784-9550
	Mount Vernon	Ad Infinitum Press	914-664-5930
	New York	The Adams Group	212-255-4900
	New York	American Pizzi Offset Corp.	212-986-1658
	New York	Andover Press	212-594-3556
	New York	Edison Lithographing	212-741-2212
	New York	Ganis & Harris	212-684-0850
	New York	Independent Printing Company	212-689-5100
	New York	Latham Process Corporation	212-966-4500
	New York	Maple-Vail Book Manufacturing	212-481-9150
	New York	Nobel Book Press	212-777-1300
	New York	Regency Graphics	212-867-5230
	New York	Waldon Press	212-691-9220
	New York	Fred Weidner & Sons Printers	212-989-1070
	New York	West Side Graphics	212-222-9304
	Plainview	Mark IV Press Ltd.	516-349-8070
	Poughkeepsie	Fairview Litho	914-473-4747
	Rensselaer	Hamilton Printing Company	518-477-9345
	Rochester	Case-Hoyt Corporation	716-232-6840
	Rochester	Tucker Printing	716-271-4570
	Rome	Canterbury Press	315-337-5900
	Spring Valley	Town House Press	914-425-2232
	Tunnel	Geryon Press	607-693-1572
	Woodside	G. Schirmer Inc.	212-784-8520
	Yorkville	Vicks Lithograph & Printing	315-736-9346
OH	Ashland	Bookmasters	800-537-6727
	Beachwood	Carnes Publications Services	216-292-7959

OH	Cincinnati	Best Impressions	800-242-9800
	Cincinnati	Johnson & Hardin Company	513-271-8874
	Cincinnati	C. J. Krehbiel Company	513-271-6035
	Cincinnati	John S. Swift -- Cincinnati	513-721-4147
	Dayton	Hooven-Dayton Corporation	513-224-1108
	Dayton	United Color Press	513-461-5150
	Lorain	Lorain Book Manufacturers	216-244-3839
	Millersburg	Mast Printing	?
	Oberlin	William Feathers / Printers	216-774-1500
	Sugarcreek	Schlasbach Printers	?
OR	Bend	Maverick Publications	503-382-6978
	Forest Grove	Times Litho	?
	Medford	Commercial Printing Company	503-773-7575
	Tigard	Germac Printing	503-639-0898
PA	Bethlehem	Oaks Printing Company	215-759-8511
	Dallas	Offset Paperback Manufacturers	717-675-5261
	Doylestown	Quixott Press	215-794-7107
	Easton	Mack Printing	215-258-9111
	Emigsville	Progressive Typographers	717-764-5908
	Ephrata	Science Press	717-733-7981
	Fairfield	Fairfield Graphics	717-642-5871
	Hanover	The Sheridan Press	800-352-2210
	Harrisburg	The McFarland Company	717-234-6235
	Harrisburg	Telegraph Press	717-234-5091
	Lancaster	Wickersham Printing Company	717-299-5731
	Lebanon	Sowers Printing Company	800-233-7028
	Mars	Publishers Choice Book Mfg.	412-625-3555
	Philadelphia	Braceland Brothers	215-492-0200
	Philadelphia	Eastern Lithographing	215-225-1150
	Philadelphia	National Publishing Company	215-732-1863
	Philadelphia	Pearl Pressman Liberty Printing	215-925-4900
	Pittsburg	Schiff Printers & Lithographer	412-441-5760
	Scranton	Haddon Craftsmen	717-348-9211
	State College	Jostens -- State College	814-237-5771
	Warminster	Neibauer Press	215-322-6200
PR	San Juan	Publishing Resources Inc.	809-724-0318
RI	Providence	Federated Lithographers	401-781-8100
SC	Columbia	R. L. Bryan Company	803-779-3560
TN	Clarksville	Jostens -- Clarksville	615-647-5211
	Collierville	Fundcraft	800-351-7822
	Kingsport	Kingsport Press	615-246-7131
	Nashville	Parthenon Press	800-251-4857
	Nashville	Rich Printing Company	615-385-3500
	Waynesboro	Southern Tennessee Publishing	615-722-5404
TX	Austin	Morgan Printing	512-459-5194
	Austin	Sweet Printing	512-255-1055
	Austin	Tracor Publications	512-929-2222
	Canyon	Staked Plains Press	806-655-1061
	Dallas	Taylor Publishing Company	214-637-2800
	Houston	D. Armstrong Company	800-231-6441
	Wolfe City	Henington Publishing Company	214-496-2226
UT	Provo	Community Press	801-225-2299

UT	Salt Lake City	Hawkes Publishing Inc.	801-262-5555
	Salt Lake City	Publishers Press	801-972-6600
VA	Fredericksburg	Bookcrafters -- Fredericksburg	703-371-3800
	Harrisonburg	Banta-Harrisonburg	703-433-2571
	Harrisonburg	R. R. Donnelley - Harrisonburg	800-428-0832
	Richmond	William Byrd Press	804-264-2711
	Richmond	Whittet & Shepperson Printers	804-649-9047
VT	Brattleboro	The Book Press	802-257-7701
	Burlington	The Lane Press	802-863-5555
	Lunenburg	Stinehour Press	802-328-2507
	Montpelier	Capital City Press	802-223-5207
	Rutland	Academy Books	800-451-6045
WA	Coulee Dam	Four Winds Press	509-633-2060
	Seattle	Dinner & Klein	206-682-2494
	Snohomish	Snohomish Publishing	206-568-4121
	Vancouver	ABC Printing	206-573-2161
WI	Madison	Fleetwood Graphics	608-829-3536
	Madison	Omnipress	800-828-0305
	Madison	Straus Printing Company	608-251-3222
	Madison	Webcrafters	800-356-8200
	Menasha	Banta Company	414-722-7771
	Muskego	Marek Lithographics Inc.	414-679-3600
	New Berlin	W. A. Krueger -- New Berlin	414-784-2000
	Stevens Point	Worzalla Publishing Company	715-344-9600
WV	Parson	McClain Printing Company	304-478-2881

CANADIAN PRINTERS SORTED BY LOCATION

BC	Vancouver	Evergreen Press	604-321-2231
	Vancouver	Mitchell Press Ltd.	604-731-5211
MB	Altona	Friesen Printers	204-324-6401
	Winnipeg	Hignell Printing Ltd.	204-783-7237
ON	Don Mills	T. H. Best Printing Company	416-447-7295
	Lindsay	John Deyell Company	705-324-6148
	Scarborough	Webcom Ltd.	416-496-1000
	Toronto	Coach House Press	416-919-2217
	Toronto	Lawson Graphics	416-251-3171
	Toronto	Ronalds Federated Ltd.	416-964-1374
	Weston	Southam Printing Ltd.	416-741-9700

PRICE COMPARISON CHARTS

The prices printers charge for manufacturing books can vary significantly from printer to printer and from book to book. In our own experience we have received quotes as low as $528 and as high as $2695 for one thousand copies of the same book -- that's well over $2.00 difference per copy. Similarly, the quotes for printing 3000 copies of the Second Edition of this Directory varied from a low of $2334 to a high of $5881 -- a difference of $3547, or over $1.18 per copy.

Why these incredible differences in prices? The differences are primarily the result of two factors: (1) the best printers tend to specialize in certain quantities, sizes, and bindings and, because they do so, they can offer lower prices for a book meeting those specifications; and (2) some printers are simply inefficient producers -- their prices are consistently high. This Directory and the following price comparison charts should help you to identify those printers who can best meet your particular book printing needs.

As you review the following charts, note that one printer might offer the lowest prices for one size book (or binding, or quantity) but not for another -- this, again, is due to their specialization. Hence, you should always query at least five to ten printers for every book you intend to publish. This simple procedure can save you hundreds of dollars on every book you publish. For example, the difference between the lowest quote we received for the printing of the Second Edition of this Directory and the tenth lowest quote was $308. So don't ignore this fundamental rule: Query at least five to ten printers every time.

Here's two more examples just to make the point: The difference between the lowest and tenth lowest quote for 1000 copies of another book was $269 -- a difference of 27 cents per copy. And the difference between the lowest and fifth lowest quote for 5000 copies of another book was $592, or about 12 cents per copy.

With those kinds of savings you can certainly afford to take a little extra time to query a few more printers. But, please, do not query indiscriminately. Do not waste your time on printers who clearly cannot provide the services you require.

Use this Directory to help you focus only on those printers who look most promising. Read the previous listings, review the following charts, consider all 17 points for selecting a book printer, and then query.

In early 1983, the Huenefeld Report surveyed 13 leading short-run book printers. As part of their survey, they asked the printers to state what their normal price would be for the production of a sample book with these specifications:

Size: 5 5/8 x 8 inches
Pages: (A) 160 pages
            (B) 320 pages
Paper: average paper quality
Cover: (1) 2-color C1S
            (2) stamped case; 2-color assembled jacket
Binding: (1) perfectbound paperback
            (2) sewn casebound

After eliminating the high and low manufacturing estimates, the average estimate for paper, printing, and binding was as follows (per copy costs):

Quantity	(1) (A)	(1) (B)	(2) (A)	(2) (B)
1500	$1.20	$1.87	$1.98	$2.73
3000	.85	1.31	1.64	2.10
4500	.75	1.11	1.48	1.92
6000	.66	1.01	1.40	1.82
9000	.59	.89	1.32	1.71

Note that the most dramatic price breaks occur almost right away, between 1500 and 3000 copies. Also notice that hardcover costs averaged about 75 cents more per copy (for 160 pages) -- regardless of quantity. The above figures should give you some basis for comparing the actual quotes you get from printers. But, please note that these price estimates were for mid-1983.

The above information is reprinted courtesy of John Huenefeld, publisher of the Huenefeld Report. This bi-weekly newsletter is an invaluable resource. We highly recommend you subscribe. For more information, write to The Huenefeld Company, P O Box U, Bedford, MA 01730. The current subscription cost is $84 -- and worth every penny.

FBP -- PRICE COMPARISON CHART

The chart on the two following pages displays the prices we obtained for 1000, 2000, and 5000 copies of one of our manuals, <u>FormAides</u> <u>for</u> <u>Successful</u> <u>Book</u> <u>Publishing</u>. The specs for the book were as follows:

Title: FormAides for Successful Book Publishing
Pages: 48 + cover
Size: 8 1/2 x 11 inches
Paper: 60 lb. white offset or equivalent
Text Ink: Black
Cover: 65 lb. color cover stock printed with
             black ink on sides 1 and 4
Binding: Saddlestitched
Packing: In tightly sealed cartons
Copy: Camera ready copy; line copy only; no bleeds

The RFQ for this book was sent out in December 1982 and January 1983. Prices, therefore, reflect costs at that time. Prices for a similar book would undoubtedly be higher now.

Here is a brief summary of the chart, listing the low, average, and high quotes for each of the three requested quantities:

	1000	2000	5000
lowest quote	$ 528	$ 814	$1672
average quote	1160	1608	2855
highest quote	2695	3270	5075
average of 10 lowest	717	1120	1985

Note the amazing differences: At 1000 copies the difference between the low and high quotes is $2167 -- well over $2.00 per copy. Even if we take the difference between the low and average quotes, the savings is $632. The two lowest bids are $528 and $540; the next lowest bids are $200 to $250 higher -- a savings of 20 cents per copy.

At 5000 copies the differences are not quite so dramatic but still significant. With a difference of $3403 between the low and high quotes, the savings amounts to 68 cents per copy. And with a difference of $1183 between the low and average quotes, the savings is 24 cents per copy. Finally, the difference between the lowest quote and the tenth lowest quote is $528 -- a nice little sum to keep in your pocket.

FBP -- Price Comparison Chart

	Printers	Quoted Prices		
		1000	2000	5000
*	Adams Press	$1110	$1791	$3300
*	Algen Press Corporation	1826	2488	4396
*	Amity Hallmark	867	1396	3061
*	Apollo Books	780	1234	2362
	Bay Port Press	1100	1543	2496
	Ovid Bell Press	1190	1590	2200
	Bireline Publishing	1049	1900	4200
	BookCrafters	913	1365	2483
	Braun-Brumfield	797	1187	2267
*	Brennan Printing	1518	----	----
	Capital City Press	1035	1343	2259
	Carnes Publications	1020	1485	2867
*	Champion Printing	1393	1835	3168
*	Copen Press	----	----	1921
	Corley Printing	780	1085	1800
	Country Press	780	1300	3250
	Delta Lithograph	1020	1418	2325
	Dinner & Klein	1150	1448	2342
	Eastern Litho	1110	1357	2112
*	Edwards Brothers	984	1410	2632
*	Franklin Press	1255	----	2988
	Futura Printing	1488	1836	2844
*	Great Northern Printing	2695	3270	5075
	Harlo Printing	1061	1213	2192

* Astericks indicate price quotes reflecting some
difference between our specs and theirs. See below:

*	Adams:	50 lb. paper / includes shipping
	Algen:	80 lb. C2S paper / 10 pt C1S cover
	Amity:	20 lb. bond paper / self-cover
	Apollo:	perfectbound / C1S cover
	Brennan:	52 pages + cover
	Champion:	self-cover
	Copen:	50 lb. paper / 67 lb. bristol cover
	Edwards:	includes shipping
	Franklin:	8 pt C1S cover
	GNP:	44 pages + cover

Printers	Quoted Prices		
	1000	2000	5000
D B Hess	----	1498	2515
Inter-Collegiate Press	970	1298	2665
Johnson Graphics	1650	2135	3580
Kingsport Press	1403	1861	3392
Kni Book Manufacturers	1122	1557	2957
Letters Inc.	1279	1910	3657
Lorell Press	1328	1684	2675
Malloy Lithographing	822	1132	1906
* McClain Printing	1491	2026	3622
McGregor & Werner	878	1482	3227
McNaughton & Gunn	1267	1593	2541
Moran Colorgraphic	1493	1882	3011
Morgan Printing	773	1364	3112
Multiprint	528	814	1672
National Reproductions	540	940	2000
Oxford Group	1080	1671	3263
* PAK Printing	723	1302	3039
Patterson Printing	910	1200	1920
Prinit Press	908	1477	2644
Print Center	1106	1832	3575
Quinn Woodbine	1277	1741	3011
Rich Printing	1391	1640	2382
Semline	1265	1619	2551
* Speedy Printers	729	1299	3089
John S. Swift Company	797	1241	2476
Thomson-Shore	780	1145	2390
Torch Publications	1781	2367	3740
* Town House Press	1140	1690	3230
Walsworth Publishing	1080	1560	2950
Whittet & Shepperson	2100	2765	4650
Worzalla Publishing	1485	1840	2845

```
* McClain:  includes shipping
  PAK:      20 lb. bond paper / includes shipping
  Speedy:   20 lb. bond paper / self-cover
  Town H.:  10 pt. C1S cover
```

DBP2 -- PRICE COMPARISON CHART

The chart on the next page displays the quoted prices we received for printing 3000 copies of the Second Edition of this Directory. The specs for the book were as follows:

Title: Directory of Short-Run Book Printers
Pages: 160 pages + cover
Size: 5 1/2 x 8 1/2 inches
Paper: 60 lb. white offset or equivalent
Text Ink: Black
Cover: 10 pt C1S, printed 2-colors, sides 1 and 4
Binding: Perfectbound
Packing: In tightly sealed cartons
Copy: Camera-ready line copy only; no bleeds

The RFQ for the Second Edition of this Directory was sent out in May 1984 and returned by late June 1984 (with a few prices received in July). Prices reflect costs at that time.

The lowest quote we received was for $2334 (for a cost of about 78 cents per copy); the highest quote was for $5881 -- that's a savings of $3547, or $1.18 per copy. The average quote for the 42 printers who responded to our RFQ was $3353, for a savings of $1019 between the average quote and the lowest quote, or 34 cents per copy.

Perhaps a better standard for judging the savings that can be obtained from querying the appropriate printers is to compare the ten lowest quotes. The difference between the lowest and the tenth lowest quotes was $308. The difference between the lowest quote and the average of the ten lowest quotes ($2479 = average) was $145.

DBP2 -- Price Comparison Chart

Printers	Shipment (work days)	Quoted Prices	
Adams & Abbott	15	$5400	
Adams Press	45-60	4500	
Apollo Books	60-75	2410	
Banta Company	?	3160	
T H Best Printing	15-20	4335	Can$
Book-Mart Press	?	2995	

Printers	Shipment (work days)	Quoted Prices	
Braun-Brumfield	20	2334	
Brennan Printing	45	4552	
Capital City Press	20-25	2417	
Corley Printing Company	15-20	2820	
Crane Duplicating	14	3485	
Cushing-Malloy	25	2642	
Delta Lithograph	15-21	3276	
Dinner & Klein	20	3810	
R R Donnelley	20-30	2787	
Edwards Brothers	22	2851	
Eerdmans Printing	15-20	2412	
Fort Orange Press	20	3960	
Futura Printing	30	3408	
Haymarket Press	23	4652	
Hignell Printing Ltd	15-20	3496	Can$
Inter-Collegiate Press	20	2831	
Jostens	10-15	2985	
Malloy Lithographing	25	2741	
Maverick Press	20	4500	
McGregor & Werner	10	2867	
McNaughton & Gunn	23	2478	
Morgan Printing	15	4214	
Murray Printing Company	?	3465	
National Reproductions	15	3916	
Patterson Printing	15-20	2469	
Prinit Press	60	2984	
Print Center	30	5881	
Publishers Press	15	2734	
Rose Printing Company	20-25	2527	
Spilman Printing	15	2527	
Thomson-Shore	25	2350	
Torch Publications	23	3840	
Vantage Press (Andover)	45	4200	
Walsworth Publishing	20	2550	

FDRM -- PRICE COMPARISON CHART

The chart below displays the quoted prices we obtained for a second printing of one of our books, <u>FormAides</u> <u>for</u> <u>Direct</u> <u>Response</u> <u>Marketing</u>. The specs for the book were as follows:

Title: FormAides for Direct Response Marketing
Pages: 80 + cover
Size: 8 1/2 x 11 inches
Paper: 70 lb. white offset or equivalent
Text Ink: Black
Cover: 10 pt C1S, printed 2-colors, sides 1 and 4
Binding: Perfectbound
Packing: In tightly sealed cartons
Copy: Reprint -- line copy only; no bleeds

The RFQ was sent out in December 1983 and returned by mid-January 1984. Prices reflect costs at that time.

Note that some printers do not stock 70 lb. white offset paper and so quoted for 60 lb. paper. The difference in cost is substantial, almost 35 cents per copy (judging from the two quotes received from Delta Lithograph). Hence, it is unfair to compare the prices of those providing 60 lb. paper versus those providing 70 lb. paper. Perhaps the most important point you can learn from this chart is that special orders cost money. Whenever possible, stick to standard papers, sizes, and bindings. It'll save you money.

FDRM -- Price Comparison Chart

| | | Quoted Prices | |
Printers	Paper	5,000	10,000
Apollo Books	60 lb	$3338	----
BookCrafters	60 lb	3605	5803
Braun-Brumfield	70 lb	3917	6899
Corley Printing	50 lb	3200	5320
Delta Lithograph	60 lb	4105	6880
Delta Lithograph	70 lb	5830	10400
Edwards Brothers	70 lb	5049	8833
Hal Leighton	70 lb	6504	13008
McNaughton & Gunn	70 lb	4032	6798
Multiprint	70 lb	4700	8500
Patterson	60 lb	3013	5186
Rich Printing	70 lb	4478	6834
Thomson-Shore	60 lb	3350	5980

ADDITIONAL PRICE COMPARISON CHARTS

The price quote comparisons on the following pages were sent to us by users of the Second Edition of the Directory. We appreciate those of you who sent in your samples. It helps to keep us up to date on prices currently being quoted by various printers.

CHART 1 -- 144-PAGE BOOK

The following quotes are for 500, 1000, and 2000 copies of a 144-page book. The RFQ was sent out and returned in June 1985. The specs are as follows:

Pages:	144 pages
Size:	5 1/2 x 8 1/2 inches
Paper:	60 lb. white offset or equivalent
Ink:	black
Cover:	10 pt C1S, 1-color, sides 1 and 4 only
Binding:	perfectbound
Copy:	camera-ready line copy only; no bleeds

	Quoted Prices			
Printers	500	1000	2000	*
Book-Mart Press	$1570	$1870	$2360	Y
Braun-Brumfield	1006	1345	2002	N
Crane Duplicating	1081	1355	2132	N
Delta Lithograph	1657	1977	2536	N
Edwards Brothers	1526	1937	2707	N
Harlo Printing	1480	1821	2549	Y
Inter-Collegiate Press	1189	1636	2183	Y
Jostens Printing	1589	1921	2570	Y
Malloy Lithographing	1149	1441	2006	Y
Maverick Press	1530	2290	3790	Y
McGregor & Werner	1325	1600	2332	N
McNaughton & Gunn	1008	1303	1903	N
Morgan Printing	1045	1590	2681	N
National Reproductions	716	1296	----	N
Publishers Press	1438	1704	2240	Y
Rose Printing	----	1722	2050	N
Thomson-Shore	1095	1360	1885	Y
Average Price Quoted	$1275	$1657	$2370	

* Y = shipping charges included;
N = shipping charges not included.

CHART 2 -- PHOTO BOOK

The following quotes are for 5000, 10,000, and 12,000 copies of a 160-page book printed on 70# enamel gloss. The RFQ was sent out and returned in May 1985. The specs are as follows:

Pages: 160 pages
Size: 8 1/2 x 11 inches
Paper: 70 lb. enamel gloss (both sides)
Ink: black
Cover: 12 pt C1S, 2-color, sides 1 and 4, laminated
Binding: perfectbound
Copy: camera-ready copy plus approximately 200
          B&W photos to be screened (150 line) and
          stripped by printer.

Printers	Quoted Prices		
	5000	10,000	12,000
Apollo Books	$ 8,093	$12,903	$14,980
BookCrafters	11,714	20,189	23,579
Colorcraft Ltd.	8,450	15,000	17,640 *
Dah Hua Printing	7,650	13,800	16,200 *
Dai Nippon Printing	11,600	19,000	21,960 *
R R Donnelley	17,649	23,629	26,020
Edwards Brothers	10,464	17,278	20,256
International Press	11,078	17,878	20,598 *
McNaughton & Gunn	14,732	------	------
Patterson Printing	17,105	30,332	35,623
Singapore National	5,200	9,500	11,160 *
Tien Wah Press	8,501	13,736	15,694 *
Toppan Printing	10,600	------	------ *
Average Price Quoted	$10,987	$17,568	$20,337

* Overseas printers -- prices in some cases include
  shipping via ship to San Francisco or Los Angeles,
  but do not include customs clearance or surface
  shipping to final destination.

Note: Many of the overseas printers offer better prices for book requiring so many halftones (a labor-intensive process). But remember that their prices may not include all shipping charges, and their delivery time is considerably longer. Most also require payment in advance by irrevocable letter of credit.

CHART 3 -- 200-PAGE BOOK (PERFECTBOUND / COMB BOUND)

The following chart displays the prices quoted for 2,500 copies of a 200-page book with two binding options (perfectbound and comb bound). The RFQ was sent out in December 1984 and returned in January 1985. The specs are as follows:

Pages: 200 pages
Size: 8 1/2 x 11 inches
Paper: 60 lb. white offset or equivalent
Ink: black
Cover: 10 pt C1S, 2-color, sides 1 and 4
Binding: perfectbound / comb bound
Copy: camera-ready line copy only; no screens

Printers	Per Copy Cost @ 2500 Copies Perfectbound	Comb Bound
Banta Company	$2.30	----
BookCrafters	1.90	$2.24 *
Book-Mart Press	2.10	2.32
Braun-Brumfield	1.80	----
Capital City Press	2.12	----
Independent Printing Co.	2.50	----
Inter-Collegiate Press	1.90	----
Malloy Lithographing	1.88	----
McGregor & Werner	2.04	----
McNaughton & Gunn	1.78	2.34
Publishers Press	2.06	----
Thomson-Shore	1.72	2.31
Town House Press	2.12	----
Average Quoted Price	$2.017	

* Comb bound prices are given only for the four lowest quotes. BookCrafters's price is for metal spiral binding rather than comb binding.

Note that comb-binding is a labor-intensive form of binding; hence, the incredible differences in prices -- as much as 59 cents more per copy. The delivery time is also longer.

Although Banta quoted one of the highest prices for 2500 copies, they offered a quote of $1.20 per copy for 10,000 copies (on their cameron-belt press).

Bibliography

These books may be ordered from Ad-Lib's Mail Order Bookstore:

Ad-Lib Consultants, _FormAides for Successful Book Publishing_
    52 pages (Ad-Lib Publications, 1983) $5.95

George Beahm, _How to Publish and Sell Your Cookbook_
    80 pages (GB Publishing, 1985) $7.95

Clifford Burke, _Printing It: A Guide to Graphic Techniques_
    127 pages (Wingbow Press, 1972) $4.95

Alastair Campbell, _The Graphic Designer's Handbook_
    192 pages (Running Press, 1983) $14.95

Peggy Glenn, _Publicity for Books and Authors_
    182 pages (Aames-Allen, 1984) $12.95

John Kremer, _The Independent Publisher's Bookshelf_, Third Edition
    32 pages (Ad-Lib Publications, 1985) $3.00

John Laing, _Do-It-Yourself Graphic Design_
    156 pages (Facts on File, 1984) $13.95

Sara Pitzer, _How to Write a Cookbook and Get It Published_
    253 pages (Writer's Digest Books, 1984) $15.95

Dan Poynter, _Business Letters for Publishers_
    82 pages (Para Publishing, 1981) $14.95

Dan Poynter, _Publishing Forms_
    portfolio (Para Publishing, 1984) $14.95

Dan Poynter, _Publishing Short-Run Books_
    100 pages (Para Publishing, 1982) $6.95

Dan Poynter, _The Self-Publishing Manual_, Third Edition
    352 pages (Para Publishing, 1984) $14.95

Marilyn and Tom Ross, _The Complete Guide to Self-Publishing_
    399 pages (Writer's Digest Books, 1985) $19.95

Russell A. Stultz, _Writing and Publishing on Your Microcomputer_
    165 pages (Wordware Publishing, 1984) $13.95

L. A. Tattan, _Publishing Yourself Without Killing Yourself_
    191 pages (InPrint, 1981) $9.95

Carl Vandermeulen, _Photography for Student Publications_
    160 pages (Middleburg Press, 1979) $12.95

INDEX

COLOPHON

Composition: produced on a Morrow MD-11 computer

    input: NewWord word processing software
        Personal Pearl database

    output: Apple LaserWriter
        with Diablo 630 emulation

    cover typesetting by Walsworth Publishing

Printed by: Walsworth Publishing Company
        Marceline, MO

Front photo: Courtesy of Publishers Press

REQUEST FOR FEEDBACK

This Third Edition of the <u>Directory</u> <u>of</u> <u>Short-Run</u> <u>Book</u> <u>Printers</u> is a direct result of the feedback we have received from the users of the first and second editions. We hope this expanded edition will better serve your needs.

In return, we ask that you help us to make any future editions even better. Please let us know what you like about this edition of the <u>Directory</u>, what more you would like to see included, the names of any book printers you know who we've left out, comments (either good or bad) about any book printers you have used -- anything, in short, which could help to ensure the accuracy and completeness of any future editions of this <u>Directory</u>.

Below are two copies of our user rating form which you may copy and complete for the book printers you have used. Rate each by circling the appropriate number (0 = worst; 5 = average; 10 = best; and other numbers for in-between ratings). Please add any additional comments about the printers on the back side of the ratings form (the comments may be quoted in a future edition of this <u>Directory</u> or any update newsletter we might issue; all comments will remain confidential and will be quoted anonymously).

RATINGS   Printer_____   How often used_____

   Address_____

Speed	0	1	2	3	4	5	6	7	8	9	10
Price	0	1	2	3	4	5	6	7	8	9	10
Dependability .	0	1	2	3	4	5	6	7	8	9	10
Service	0	1	2	3	4	5	6	7	8	9	10
Quality	0	1	2	3	4	5	6	7	8	9	10
Overall	0	1	2	3	4	5	6	7	8	9	10

RATINGS   Printer_____   How often used_____

   Address_____

Speed	0	1	2	3	4	5	6	7	8	9	10
Price	0	1	2	3	4	5	6	7	8	9	10
Dependability .	0	1	2	3	4	5	6	7	8	9	10
Service	0	1	2	3	4	5	6	7	8	9	10
Quality	0	1	2	3	4	5	6	7	8	9	10
Overall	0	1	2	3	4	5	6	7	8	9	10

DELUXE MAIL MERGE EDITION NOW AVAILABLE

NO MORE HAND ADDRESSING! NO MORE SIFTING THROUGH THE BOOK!

Well, you asked for it, so we're offering it. The addresses of all printers now listed in the Third Edition of the Directory of Short-Run Book Printers are now available in mail merge format for your convenience in sending out Requests for Quotations. These address files are compatible with the WordStar and NewWord word processing programs. Here's what is included:

1) a standard Request for Quotation form already formatted for mail merging;

2) complete lists of all book printers listed in the Third Edition of the Directory of Short-Run Book Printers sorted in two different ways: alphabetically and by zip code;

3) 22 different lists of short-run book printers sorted by various classifications, including:

ultra-short-run specialists
short-run specialists
self-publishing specialists
typesetting printers
teletypesetting printers
hardcovers (in-house)
4-color printers
advertising printers
Canadian book printers
overseas book printers
Top 15 short-run book printers

annual reports
catalogs
cookbooks
demand printers
galley copies
journals
magazines
mass-market paperbacks
juvenile picture books
yearbooks
booklets

4) label formats for 1-up, 2-up, or 3-up printing.

This Deluxe Mail Merge Edition is now available for over 75 computer formats via the Uniform software program. Here are some of the computer disk formats (SS or DS) included:

Columbia 964
DEC VT-180
Epson QX-10
Fujitsu 16S
Heath CP/M
HP-125

IBM-PC CP/M 86
IBM-PC PC-DOS
Kaypro II
Morrow CP/M
NCR Decision
NEC PC-8000

Olivetti
Osborne 1
Otrona Attache
Phillips
Sanyo MBC-1000
Chameleon CP/M

Televideo
TI Professional
CP/M 86
Wangwriter
Xerox
Zenith

This Deluxe Mail Merge Edition is now available for only $25.00 if you already have the book edition of the Directory, or for $30.00 if you also want a copy of the Third Edition of the Directory. Please let us know what disk format you require when you order. See order form on the next page. Thank you.

239

DID YOU BORROW THIS COPY?

Have you been borrowing a copy of this <u>Directory</u> from a friend or colleague? Don't you wish you had your own copy for quick, easy reference? To make it easy for you to order, we've included the simple order form below.

_____ $12.00 **Directory of Short-Run Book Printers, 3rd Edition**

_____ $30.00 **Deluxe Mail Merge Edition** - format:_____
(with book)

_____ $25.00 **Deluxe Mail Merge Edition** - format:_____
(without book)

_____ $ 5.95 **FormAides for Successful Book Publishing**

_____ $ 9.95 **FormAides for Direct Response Marketing**

_____ $ 3.00 **The Independent Publisher's Bookshelf, 3rd Ed**

_____ $10.00 **Book Marketing Made Easier** (available Feb 86)

_____ $19.95 **101 Ways to Market Your Books** (available May 86)
(hardcover)

_____ $14.95 **101 Ways to Market Your Books** (available May 86)
(softcover)

_____ $_____ other books:_____

_____ $_____ _____
======= ======

_____ _____ Total for Books

_____ _____ Postage & Handling ($1.00 per book)

_____ _____ TOTAL

Please rush me the books I've checked above. It is my understanding that if I am not completely satisfied with any book, I may return it within 15 days for a full refund. Thank you!

Name_____ Phone_____

Address_____

_____

For faster delivery, use your charge card -- call (515) 472-6617

Mail To: Ad-Lib Publications, P.O. Box 1102, Fairfield, IA 52556-1102